Agribusiness Start-up Toolkit

THE AUTHOR

Naveed Hamid possesses a meritorious academic record and awarded by university merit certificate in recognition of his having achieved the prescribed scholastic standards of high order OGPA in Doctorate as well as in Masters (Agri-Business Management) programme. He has more than 6 years of teaching and research experience in the field of Agricultural Business. He has more than 85 research publications in referred, NAAS rated and UGC approved journals, 22 abstracts, 31 Book chapters,8 Agribusiness Case Studies, several Newspaper/Magazine articles, 17 books and so many international/National conferences to his credit. He is recipient of more than 14 International/National recognition award from repute organization for his contribution to the Agri-Business field. Among those awards he is bestowed with Swami Vivekananda Excellence Award by Governor of Kolkata, Krishi Bhushan, Best Mentor Award in Innovation & Entrepreneurship, Best Idea Award for Smart Fish, and IRDO Young Fellow award. He has Mentored 100 + Students/Faculties in Agriculture and allied sectors. His field of expertise is Agricultural Business Management, Supply chain/value chain and marketing. Naveed Hamid is presently working in Innovation & Entrepreneurship Cell IDP SKUAST Kashmir and has worked in Ministry of Agriculture Cooperation and Farmers Welfare. He is also Ambassador, ACPCL (M.P) for promotion of FPO and Agribusiness related activities in Jammu and Kashmir State, District Coordinator, IMPAR, Advisor,MKPA international Nodal Officer Innovians Pvt Limited, Coordinator Centre of Excellence for Dairy (CEDSI) and Peace Ambassador, Nepal.

Agribusiness Start-up Toolkit

Take your e-mentor with you

Naveed Hamid
Business Development Analyst
Innovation & Entrepreneurship Cell SKUASTK

2022
Daya Publishing House®
A Division of
Astral International Pvt. Ltd.
New Delhi – 110 002

ISBN: 978-93-5461-087-5 (HB)

Published by : **Daya Publishing House®**
A Division of
Astral International Pvt. Ltd.
– ISO 9001:2015 Certified Company –
4736/23, Ansari Road, Darya Ganj
New Delhi-110 002
Ph. 011-43549197, 8130496929
E-mail: info@astralint.com
Website: www.astralint.com

Digitally Printed at : **Replika Press Pvt. Ltd.**

Foreword

Entrepreneurship is an attitude, a complete philosophy, a manner of thinking, and a way of life. An entrepreneur looks at a problem and wants to solve it in a new and interesting way. He/she is a creator in the truest possible sense.

India's next wave of growth will be driven by the increasing number of new age entrepreneurs. From expanding traditional family run businesses to setting up unicorns, the Indian start-up ecosystem has never had it so good.

India has no dearth of talent, what we need is to channelize their ideas and efforts into ventures that have the potential to grow and generate wealth and employment. We also require people who will combine business goals with a social purpose.

I am happy that Naveed Hamid, BDA, Innovation & Entrepreneurship Cell IDP SKUAST K has pioneered and conceptualized the first-of-its-kind, Agribusiness Start-up Toolkit that aspires to guide and support promising innovators in the path of Innovation & Entrepreneurship.

This toolkit presents fundamental insights into Agricultural entrepreneurship, Start-ups and guide for young budding entrepreneurs and innovators that are missing from much of the recent literature on the subject.

The distinctiveness of the Toolkit lies in the rigor of its practice component and experiential focus that guides participants from design thinking, ideation, Business Plan, funding options, company registration, pitching to investors etc to venture creation.

The main mission is to give wings to entrepreneurial dreams and ideas through this Toolkit. I congratulate him for his dedication and utter hard work to finalize this ever-needed book in the times of change.

Prof. Nazir Ahmad Ganai
Vice-Chancellor
SKUAST Kashmir

Foreword

Agriculture plays a significant role in the Indian economy. Almost 70% of Indian agriculture is used to support the livelihood of 70%. While the onset of Covid-19 disrupted several industries, the agritech sector has proved to be surprisingly resilient. With the Aatmnibhar Bharat Initiative, the network of critical stakeholders for continuous scientific and policy advice, research partnership among startups & educational institutes; the country is witnessing the commercialization of new agri-technologies. Agribusiness is playing an essential role in revitalizing the agricultural sector in India as it generates diversified income, providing widespread employment and entrepreneurial opportunities in rural areas.

With 43 startups in 2013 to more than 1,377 startups today, the Agritech sector in India has come a long way, motivated by increased digitization, government initiatives, and growing investor's interest in this domain. Startups like **Stellapps**, which is one stop dairy supply chain digitization via IoT. **Ninjacart** is a B2C and B2B agritech startup that solves the fresh agricultural produce supply chain problems for farmers and retailers. **CropIn Technology** provides a one-stop SaaS-based solutions platform enabling agribusinesses to increase efficiency, productivity, and sustainability. Another platform, **DeHaat** is a technology-led platform that offers end-to-end agriculture services to Indian farmers with an aim to improve the efficiencies of the agriculture value chain.

With the advancements in technology, the Agritech sector will not only reach new heights but also with the rise in technology adoption among farmers and increased opportunities in smart farming, precision farming, robotics & automation, and vertical farming, there is no doubt that Agritech sector will grow explosively in the very near future.

This book is a compilation on how to pursue agripreneurship in India, which will be helpful for the younger generation, especially to those who are living in remote regions of the country, and will also enhance the awareness of the stakeholders such as State Agricultural Universities, Banks, and KVKs about the business opportunities in the agriculture sector. This book shall serve as a guiding light for the academicians and fresh graduates aspiring to become entrepreneurs.

Prof Amitabha Bandyopadhyay

Foundation for Innovation & Research in Science & Technology (FIRST)

Indian Institute of Technology, Kanpur / भारतीय प्रौद्योगिकी संस्थान, कानपुर

Kanpur, Uttar Pradesh - 208016

Foreword

Entrepreneurship is an area of prime focus by the Government, private organisations, and educational institutions these days. It is becoming an integral part of professional and vocational training and gaining momentum day by day. There is a need for making available appropriate subject matter giving complete exposure on the subject to the emerging entrepreneurs for ideation, conceptualisation, implementation and running of an enterprise.

This '**Agribusiness Start-up Toolkit**' written by Sh. Naveed Hamid is a step towards this direction providing complete information for the entrepreneur. The effort is first of its kind in the agricultural arena, aiming at sensitization of young budding entrepreneurs and innovators in the agricultural sector, how they can change their simple ideas into a profitable Entrepreneurship/Start-up model. The author has covered every aspect from ideation to incubation to funding to schemes to mentoring to business model canvas to registration of the start-ups or companies etc.

Author of this book is an Agriculture Business Technocrat presently working in innovation and Entrepreneurship cell of Sher-e-Kashmir University of Agricultural Sciences and Technology Kashmir. He possesses a vast experience in Entrepreneurship, start up, mentoring and handholding. Presently he is Nodal officer and coordinator of many leading incubation and Entrepreneurship centres across the country.

This toolkit is a ready reckoner for the reader covering all aspects of entrepreneurship path. This will help the student cum faculty innovators and budding entrepreneurs in simplifying their road from idea to start-up and establishment of an enterprise.

I hope this toolkit will help promote self-employment among youth of the country at this time when opportunities for direct employment are limited.

Jaswant Singh Kalsi
Chief operating officer
Centre of Excellence for Dairy Skills in India

Acknowledgment

I Express my gratitude to **Prof. Nazir Ahmad Ganai**, Vice Chancellor, SKUAST Kashmir and Principal Investigator IDP NAHEP of SKUAST Kashmir for giving me an awesome opportunity to work independently as the BDA that has helped me to translate my knowledge into practice of producing young student Start-ups in the University.

I am dedicating this toolkit to **Prof. Nazir Ahmad Ganai**, Vice Chancellor Sher-e- Kashmir University of Agricultural Sciences & Technology of Kashmir, Team IDP and all innovators of SKUAST Kashmir. Their timely support and guidance motivated me to compile this First of its kind Agribusiness Toolkit for benefit of young budding Agriprenuers/Innovators/Start-ups.

Acknowledgement

[illegible]

[illegible]

Let's Taste the recipe of Entrepreneurship

Entrepreneurship is defined as the willingness to take risks, plan, organise, streamline, and manage a new business idea in a fast-paced global market. Entrepreneurs are self-motivated, creative individuals who are on a mission to create one-of-a-kind goods and services. Entrepreneurship is a challenging path to take, but it is also a rewarding one.

Only a generation ago, Indians solely focused on well-defined and tested career paths. Young adults usually followed the footsteps of their parents and opted for a secure career. Today, entrepreneurship has become one of the significant career choices being pursued by the youth because of the following reasons:

- The desire to build and implement a company plan from the ground up.
- A desire to be their own boss and in charge of their own destiny.
- The desire to take chances in a changing global economy with numerous untapped prospects.
- The goal is to be financially successful and to accelerate their expansion to astonishing new heights.

While most of us have the talent and desire to begin something new, it needs to be understood that successful entrepreneurship is not just about pursuing novelty. It is about ideation that helps solve a pertinent issue and has a market and a customer the potential to be sold for profit. If you have plans for joining the start-up ecosystem, perhaps you should ponder over these questions, first:

- Do you have a unique idea/concept for starting up?
- Will you be able to develop a 'people and product' network to facilitate your business objectives?

- Have you developed clear and defined processes that encompass the development, promotion and delivery of the product or service?
- Have you provisioned for funds sufficient enough for your idea to take off and ensure business sustainability until revenue generation?

Being an entrepreneur instils life skills, fosters creativity and problem-solving abilities, improves cooperation, communication, networking, and fosters a never-say-die mentality, as well as a better understanding of markets and economics. Failing in entrepreneurship is a valuable lesson in and of itself. It is unrealistic to believe that every entrepreneur becomes wealthy, but if one learns to endure and overcome failure, entrepreneurship can open the door to virtually endless earning potential.

Entrepreneurial talents are applicable to practically all aspects of life and occupations, and they are not confined to business. Time management, delegating, tenacity, critical analysis, and a variety of other talents become second nature to entrepreneurs.

Today, industrialised economies are shifting from being 'managerial' to being 'entrepreneurial.' As a growing country, India is witnessing unprecedented development in entrepreneurship and is ensuring that young people embrace entrepreneurship as a career option. Being a first-time entrepreneur, a serial entrepreneur, an angel investor, a partner in a VC or PE firm, a corporate entrepreneur, an educator, or a mentor are all examples of entrepreneurial careers.

The current ecosystem for entrepreneurship is vibrant. Entrepreneurship is, therefore, a great way for the young generation to explore their areas of interest and career orientation.

Contents

Chapter-1

Entrepreneur & Entrepreneurship

Start-up Toolkit?

An entrepreneur start-up toolkit is a collection of educational tools designed to assist innovators in learning the fundamentals of entrepreneurship and increasing their chances of success. This can equip you with a solid understanding of how to run your firm as well as practical tools.

This toolkit is a step-by-step guide for emerging or establishing businesses in our communities, and it may be looked on as a blueprint for your business that helps you organize your thoughts. Nobody enjoys being trapped on a blank page with a wonderful idea but no idea where to start.

The entrepreneurial path is frequently a series of juggling acts, each of which necessitates a certain amount of knowledge in a variety of disciplines. The purpose of this toolkit is to guide the Innovators/entrepreneur through these areas and provide resources to assist them.

You have an idea, the determination to make it happen, but you don't know where to go next. As an entrepreneur, you want to do it all. But you might need some help. Startup Toolkit is a roadmap and guide for young budding innovators in the path of entrepreneurship.

The Key to Success

You may set yourself up for success by utilising the resources available to you. Your newly acquired knowledge and skills will improve your chances of founding a long-lasting business that will allow you to accomplish what you love for a living. Gaining an understanding of what works in business can help you develop your skills and tap into your own, as well as your country's or organizations, strengths and abilities. This will help you succeed in every facet of your business.

Who is an Entrepreneur?

An entrepreneur is an individual who creates a new business, bearing most of the risks and enjoying most of the rewards. The entrepreneur is commonly seen as an innovator, a source of new ideas, goods, services, and business/or procedures

The characteristics of a successful entrepreneur

- Several common features those flourishing agricultural entrepreneurs have:
- Initiative and capability to identify and take the proper benefits of the opportunities;
- Ought to be single-minded and willingness to adopt innovative as well as traditional opinions;
- Should be an inventive leader and ready the disturb the quo of the status;
- Always looking for some creative ideas to make improvements and expand their organization;
- Operate a business in a different zone that takes more ideas that are less popular in other's perspective and opinion;
- Combat in a good way with the threat as well as insecurity;
- Ought to be visionary and evaluate the risks to calculate it before;
- Organizing people in a good way and build up strategies as well as technologies to suits with the changing environment;
- Should be a creative problem solver and understand the whole process of decision-making.

What is Entrepreneurship?

The ability and willingness to take on the invention, organisation, and management of a profitable endeavour, with all the dangers it entails, in exchange for a profit.

Although entrepreneurship is commonly viewed as vital for economic development, it is not explicitly addressed in the most influential economic growth theories or development assistance programmes. Entrepreneurship is difficult to quantify.

Entrepreneurship isn't necessarily a reaction to a business opportunity (Opportunity entrepreneurship" and "necessity entrepreneurship). People commonly start micro and small firms because there are few other income-earning opportunities in places where labour supply substantially exceeds the number of paid employment available. They become entrepreneurs as a result of necessity rather than choice. An important distinction to note is that businesses founded by opportunity entrepreneurs have the ability to expand and hire additional employees, but organisations founded out of necessity are unlikely to do so. This toolkit is mostly focused on opportunity entrepreneurship.

Entrepreneurship is commonly measured by the number of self-employed persons; however, this can be misleading because many self-employed people become entrepreneurs out of necessity rather than opportunity. In any case, a country's number of entrepreneurs is merely an approximate indicator of the strength of its entrepreneurial ecosystem. The quality of its entrepreneurs is likewise critical, and even more difficult to assess.

Sure, you're an entrepreneur. But which one? Need-based entrepreneurs are those who start micro or tiny businesses in order to support themselves. Traditionally, these are unstructured, family-owned businesses that sell modest handcrafted items or meals near their homes; they rarely employ others. High-growth entrepreneurs are individuals who, in the past, formalised their enterprises out of need. They earn money, hire people, and, over time, contribute to the expansion of a sector by introducing a new product to the market.

Agricultural Entrepreneurship

Agricultural Entrepreneurship, also known as Agripreneurship, means the term which is associated with the marketing as well as manufacturing of different agricultural products and inputs too.

Small farmers produce food for their family but at the same token, they sell a small portion of their products into several markets and it increases the level of market sales. So, the working of the small farmers indicates that they are an agricultural entrepreneur. In this way, the Entrepreneurs can work solely and keep the profit from the manufacture or they can be a part of Farmer's group and sell the crop collectively and get the profit accordingly. Moreover, the farmers enter into the business relationships with other value chain partners such as agents, and also, this forms an approach of contractual marketing which may lead to large business industries at the end if they get success.

Scalable start-up entrepreneurship

Scalable start-up entrepreneurship can be defined as a business model where an organization is started on the basis of a unique idea. It involved everything that ranges from creating a plan to all the way up launching the business. This form of Entrepreneurship involved a great sort of work, education, and experience.

This type of entrepreneurship commences with a vision that they can make some changes in the world. Their investment comes from the venture capitalist as well as by employing efficient employees. Their goal is to find out the repeatable and scalable business model. Once they find that model, they further started raising funds from venture capitalists for the growth of the business. A small proportion of all trades are made up of scalable start-ups because of the risk capital and outsize return. Face book, Instagram and online shopping for electronics are the few examples of the scalable start-up business.

Importance and Role of Entrepreneurship

Entrepreneurship is the basic pillar of economic and industrial development of a country. Entrepreneurs by setting up new enterprises discover new markets and offer innovative products. They fulfil the wants and needs of people and improve their standard of living. According to Bolinger, "Creating of new business venture is the key for technical and economic development in modern market-based economies. Through entrepreneurship, people can lead a better, long and more prosperous life.

Role of Entrepreneurship Development in times of change

There is vast role of entrepreneurship in present competitive arena in transformation and ushering the time of innovation and start-ups viz:

Economic development of nation: Entrepreneurship is the basic pillar of economic development of any country. Entrepreneur creates employment opportunities in the country by making most productive use of resources. It contributes to national income which in turn, strengthens the rate of capital formation and leads to country's rapid economic growth and development.

Helps in developing new products and techniques: By entrepreneurship, new products and techniques are developed, transactions, utilities and markets are created that contribute some value to a community and marketplace. Entrepreneurs keep innovating something new and better.

Creation of employment opportunity: Entrepreneurship involves exploring new opportunities, creating new ventures, new jobs, new tastes, promoting innovation, introduction of something new and better which automatically helps in creating job opportunities.

Opportunities to exploit full human potential: Entrepreneurs use their creative efforts, vision and courage to make their business successful. For growth of business, they work with full physical, mental and intellectual capacity. In this way they make possible the full exploitation of human potential.

Promotes capital formation: Entrepreneurship is an effort to mobilize the idle funds for the establishment of new venture for their expansion and growth. An entrepreneur invests his own savings and savings from public through shares and debentures as capital to start an enterprise and for its smooth running.

Balanced economic development: Entrepreneurs help to remove regional disparities through setting up of industries and business in less developed areas and backward areas. Growth of industries and business in these areas leads to a large number of public benefits like road transport, economic and school infrastructures.

Encourages research and investigation: Entrepreneur emphasizes novelty. This can be related to product or technique of marketing. To maintain their innovation and inventions, entrepreneurs have to continuously do research and investigation.

Helps in social change: Entrepreneurs with their unique vision, leadership and offerings of new goods and services, can create many social changes and can reduce dependence on obsolete systems and technologies. They can solve social problems like unemployment by creating jobs. They increase per capital income which in turn increases the standard of living.

Relationship between entrepreneurship and economic development

Economic development can be defined as the population's economic well-being improving over time. This definition includes not just economic growth (as measured by gains in GDP per capita), but also substantial changes in the economy's structure. As a result of these developments, the focus of national output has shifted from agriculture and other primary activities to industry and, eventually, services. Modern technologies are increasingly being used in the process of structural re-alignment. Entrepreneurs are mostly responsible for bringing about fundamental or systemic changes in the economy that can be crucial for economic progress.

Entrepreneurship promotes economic growth by boosting income and the number of well-paid jobs accessible. Most members of the labour force in low-income economies are self-employed or unpaid family workers who work in agriculture or extremely small scale industrial or service companies. As the economy develops, the relative importance of various types of jobs decreases. As a result, the creation of additional paid jobs is directly related to structural and economic improvement. Economic development, as Adam Smith highlighted more than three centuries ago, entails expanding markets and growing labour division.

Entrepreneurships as a boom to Development and Prosperity of Nation

Nations develop through a combination of successful entrepreneurship Nations grow as a result of a mix of successful entrepreneurship and the power of existing companies. At varying levels of GDP per capita, however, the composition varies. Self-employment creates jobs and marketplaces at low levels of national income. However, as GDP per capita rises, larger enterprises are able to meet the expanding demand of growing markets and play a larger role in the economy as a result of the adoption of new technologies and growing economies of scale. Rising income levels provide up new business opportunities and provide more people with the tools they need to take advantage of them. Inefficient marketplaces give chances for vigilant entrepreneurs in low-income countries; in richer and more dynamic economies, more inventive innovation will be required.

Innovation, or the adoption of new technologies, is a critical entrepreneurial activity that, in industrialised countries, typically entails the use of basic science and engineering to generate wholly new goods and processes. Transferring existing technology from more developed countries and adapting them to local markets and situations is likely to be involved in emerging countries.

Entrepreneurship's growth has enormous potential to benefit both society as a whole and the individual entrepreneur. It has the potential to enhance employment creation, resulting in higher earnings and improved wellbeing for members of society who have little to sell other than their own labour. There is a justification for government and donor funding because of these external benefits. This support should not just be focused on quantity, such as the number of entrepreneurs, but also on quality, such as the type of impact they have on society (for example, "high-impact" entrepreneurs).

Quality not Quantity

Particular attention should be paid to the quality of entrepreneur in efforts to promote entrepreneurship that ultimately stimulates broad-based economic growth, meaning those who might end up hiring large numbers of people, introducing a new product or service into a sector, or generating substantial revenue such that incomes and business flow increase, and thus positively impact the economy.

Can Entrepreneurship be promoted?

The ability and extent to which players in both the public and commercial sectors can foster entrepreneurship is a hotly disputed topic among professionals in the subject. The role of government is realistically restricted to building a suitable institutional framework, according to those who favour a lone macroeconomic development approach. In practise, this means fostering a business-friendly climate, which includes issues like property rights, contract enforcement, good infrastructure, and reducing cumbersome laws, taxes, and corruption. While these issues are essential, another viewpoint is that a number of initiatives, including the role of academia and training institutions, may and should be used to promote entrepreneurship.

Entrepreneurship will develop if three key conditions are present.

- **First,** there must be opportunity for new enterprises to exist and for entrepreneurs to succeed through their own efforts.
- **Second**, there must be a propensity to enterprise, involving such characteristics as an urge for excellence, a willingness to bear risk, and a desire to be independent.
- **Third**, the ability to enterprise must be present; meaning the technical and business capabilities required to start and manage a business must exist.

Approach to increasing entrepreneurial opportunity

Promoting the image of successful entrepreneurs as role models is one strategy to assist in bringing about such transformation. Additionally, attempts to convey the real benefits of entrepreneurship, such as increased job creation, are critical in promoting the field's good image. Governments and other development players, on the other hand, would find it impossible to dramatically alter cultural norms due to the time and money required.

A multi-pronged strategy that promotes the rule of law and enhances the private sector, as well as civil society in general, can help to strengthen entrepreneurship in many parts of the world. Educating the public on the importance of pro-business changes could aid in the formation of consensus and the improvement of a country's or regions economic standing.

Chapter-2

Design Thinking
(Polish your Mind with the cream of ideas)

Design thinking is a method of problem-solving that is both creative and innovative. It's crucial for entrepreneurs who wish to come up with new and creative concepts. After all, every great business is built on a foundation of genuine understanding of people and how to effectively serve them. The design thinking method allows us to perceive our company from the perspective of the client. It teaches how to identify the customer's wants and needs, as well as how to come up with creative solutions to meet those demands.

Ideation is to combine the understanding of the problem space and people we are designing for with our imagination to generate solution concepts. Ideation is about pushing for a widest possible range of ideas from which we can select, not simply finding a single, best solution.

Preparation before the Storm

When we're ready to embark on a journey into stormy area, we know we need to be prepared for a bumpy ride and take extra care so we come out on the other side and arrive at our destination unscathed. It's the same with an ideation process like Brainstorming. Going into a Brainstorm without any preparation is a recipe for disaster. This could harm team's perception of ideation unwittingly, scaring them away from future creative endeavours. This can damage team cohesion by going into a situation like this and causing team members to fall out with each other due to a brainstorming session gone wrong.

How to Prepare Before You Start Ideating

Even though Design Thinking is not a linear process, it is critical to consider the first two stages or modes before beginning to brainstorm. We risk being lost if we fail to consider these two modes and their guidelines prior to an Ideation session. The Empathise and Define principles will assist you in gaining enough background knowledge and establishing a clear aim for your ideation sessions.

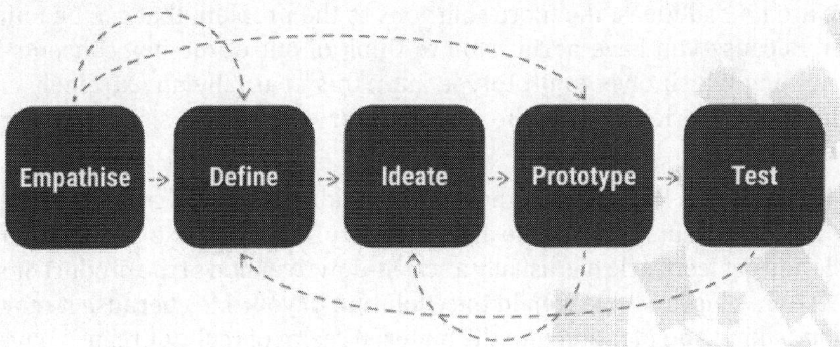

Process of Design Thinking

The design thinking process is divided into five parts. As an innovation strategy, it combines creative thinking, logical reasoning, and testing. The process' main purpose is to provide a finished product or service that meets the needs of end users. To better connect with customers, refine the emphasis of your products, and develop the business, implement the methods effectively.

1. Observe & empathize: Curiosity is at the heart of the design thinking process. Observing individuals and their wants and needs is the first step in developing successful ideas. What causes people to act the way they do? Why is it that things are the way they are? What works well, what doesn't work well, and why? Asking these questions with an open and curious mind will help you gain a better understanding of the world around you and how you might contribute effectively to it. You'll be able to notice the finite details and potential for creative issue solving that exist inside them. This fosters empathy, which is an important component of the design thinking process. Empathy will assist you in identifying the greatest product and service concepts to meet your clients' demands. Observe as much as possible during this phase. Everything you gather is useful information that will inform the future work you create.

2. Define the problem: Reflect on the information gathered from observations. Doing so will help you gain clarity about the tangible problem you're trying to solve. Group and cluster ideas together until you find the prominent themes. Then synthesize the information to help you pinpoint the most significant problem that needs to be addressed. You can't solve all customer problems. Concentrate on the most significant or impactful issue as your central focus to move forward with. This step is crucial because it centers your energy and instructs your future designs. The end goal is to convert the defined problem into a tangible, human-centered statement, rather than focusing on technology, monetary returns, or specifics of a product.

3. Ideate solutions: Now that you have a better understanding of the target audience and the problem, it's time to come up with solutions. To come up with fresh answers, think beyond the box. Brainstorm! Make an effort to be inventive. There are no bad ideas; the more solutions to the problem that can be found, the better. Because you have permission to think of out-of-the-box concepts before determining which ones to prototype later, this is usually an extremely creative and liberating period. This is the era that everyone adores since it has limitless possibilities.

4. Prototype: Phase four of the design thinking process is creating a prototype. Take your top idea(s) and create a simulation of the design so you can show it to people and get feedback. It is usually a scaled-down version of the product or system in question. This is a huge step in the evolution of your idea because it moves you beyond talking and ideation into the material realm of reality. Create a physical or digital prototype of some aspect of your product, service, or experience that you can show to users in your target market. Through trial-and-error, you will identify which of the possible solutions is best suited to solve the problem.

5. Test: All of efforts come together in the final round of product testing. This is the time to show prototype to potential customers and solicit comments. As a result, this is still a participatory stage. As you obtain a deeper knowledge of the consumer, testing a product typically leads to altering and redefining problems and solutions. It's a crucial opportunity to ensure that your entire concept is centered on the individuals who will be utilizing it. You want to know their opinions, both positive and bad. Use the data to fine-tune and flush out all of the specifics of your design. That's how you make the best product you can and get it ready to market.

Need of design thinking by Entrepreneurs/Innovators?

Focus, investment, and, most importantly, methodologies are required for thinking out the box and coming up with fresh ideas and design thinking will help in doing so. Under design thinking, risk variables will be calculated as well. It's a assured way to success, and it can help any entrepreneur stay connected and current in a constantly changing market.

1. Long-term strategy planning: Issues that are complex, hard to define, unique, and have unknown consequences need to be defined. Problems such as poverty eradication, sustainability, healthcare, and education are examples of wicked problems. As an entrepreneur, it is important as part of your long-term planning and business strategy to consider these aspects as well. In the design thinking process, we often ask questions about how one might be able to solve these problems in the future. It encourages innovation and helps in the ideation session. Taking out time to set goals and how to overcome any shortcomings is key to becoming a successful entrepreneur.

2. Allows divergent thinking: Design thinking follows an approach that allows divergent and convergent thinking. This is typically associated with imaginative thinking, originality, curiosity, and exploring various possible outcomes. Convergent thinking revolves around bringing these ideas together to come up with a single solution. Design thinking allows you, as an entrepreneur, to follow different thinking styles and explore open-ended options to come up with an actionable solution. Design thinking tools such as empathy are great for this. Prototyping also plays an integral role in this aspect.

3. Incorporate a process model: Incorporating a design thinking process can help you track and measure your progress and outcomes. It also allows you to evaluate feedback from various entrepreneurial activities. There are multiple ways to do so, but the design thinking process follows a systematic process that consists of five steps. Empathize, define, ideate, prototype, and test. Following this process as an entrepreneur can help with creating a framework and timeline to be followed.

4. Be curious and observe: Design thinking is all about observation. Being curious as to how things are, why they are a certain way, and trying to come up with solutions to this curiosity is what makes you as an entrepreneur very unique. It will eventually build empathy in you and that is one of the key aspects of design thinking.

Point of View Template – Example

User	Need	Insight
An adult person who lives in a city	To use a car for 10-60 minute trips 1-4 times per week	The user would not want to own his own car as it would be too expensive compared to his needs. He would like to share a car with others who have similar needs, however, there are no easy and affordable solutions for him. It's important for the user to think and live green and to not own more than he truly needs.

Point of View Template

User	Need	Insight

Sample Template for Design thinking

Chapter-3

Ideation

Ideation is a broad term that encompasses the entire creative process of generating and expressing new ideas. Ideation is creative thinking that aims to solve a problem or provide a more efficient way to do or accomplish something. It includes coming up with new ideas, developing current ones, and determining how to put new ideas into effect.

Ideation is the third phase of the Design Thinking process, and it's all about generating ideas. Before we explore ideation in more detail, let's briefly recap on the five stages of Design Thinking: Empathise, Define, Ideate, Prototype, and Test.

In the business world, ideation refers to things like inventing and/or developing new products or services, as well as establishing new manufacturing or distribution methods for those products or services.

Aim of an Ideation:

To think outside the box by discovering and exploring new viewpoints and approaches. It is critical that the ideation phase be a "judgment free zone" for the sake of invention and creativity.

From learning about people and the problem to coming up with solutions, the ideation phase is a critical stage. Ideation, when done correctly, is where creativity thrives; it should aid in the discovery of that game-changing answer that users have been lacking!

Ideation Methods

- Creative pause
- Cheat storming

- Crowd storming
- Daydreaming
- Provocation
- Forced relationships
- Roleplay
- Visualization
- Wishing
- Sketching and sketch storming
- Synectic's

Ideation Working Plan

Ideation may present itself in any one of a wide variety of ways and arenas.

1. **Solving Problems:** Ideation is often specifically aimed at problem-solving. For example, production managers at a company may be charged with coming up with ideas on how to reduce production costs.
2. **Derivative Ideation:** Derivative ideation refers to building on an existing idea, such as developing complementary products or accessories to sell along with a company's main product.
3. **Innovation:** An example of innovation ideation is the process of a pharmaceutical company developing new medicines. Such a type of ideation often involves doing extensive research and experimentation as part of the ideation process.
4. **Development of a "Revolutionary Idea:** Ideation sometimes ends up creating a totally new line of thought or set of ideas, such as the development of a new philosophy.
5. **Serendipitous Ideation:** Serendipitous ideation refers to situations where someone just happens to come up with a new idea even though they weren't consciously trying to do so.
6. **Combination Ideation:** Ideation often includes combining multiple ideas to create a new process or way of doing something.

Ideation and Brainstorming

Ideation is frequently associated with brainstorming, a strategy for generating new ideas. The main distinction between ideation and brainstorming is that ideation is typically thought of as an individual endeavor, whereas brainstorming is nearly always a group activity. Brainstorming is usually done by gathering a group of people to come up with either broad new ideas or particular ideas for fixing a problem or dealing with a situation.

Participants in a brainstorming session are encouraged to freely toss out whatever ideas may occur to them. The thinking is that by generating a large number of ideas, the brainstorming group is likely to come up with a suitable solution for whatever issue they are addressing.

The lines between ideation and brainstorming have become a bit more blurred with the development of several brainstorming software programs, such as Bright idea and Idea wake. These software programs are designed to encourage employees of companies to generate new ideas for improving the companies' operations and, ultimately, bottom-line profitability.

The goal of Ideation in a Design Thinking project is to produce a huge number of ideas that the team can then filter and cut down into the best, most practical or most original ones in order to inspire new and better design solutions and products.

The phase of the design process in which you concentrate on idea generating is known as ideation. In terms of conceptions and outcomes, it symbolizes a mental process of 'going broad.' Ideation is both the source material and the fuel for prototyping and bringing creative ideas into the hands of your people.

Ideation Will Help

1. Innovate with a strong focus on your users, their requirements, and your knowledge of them by asking the correct questions.
2. Go beyond the apparent solutions to boost your solution's innovation potential.
3. Bring your team's opinions and abilities together.
4. Find new ways to innovate in unexpected places.
5. Increase the quantity and diversity of your innovation possibilities.
6. Get the obvious solutions out of your head and push your team to think beyond the box.

Why to Ideate

The main objective of ideation is idea generation, or to create as many ideas as possible. This allows your team to transition from identifying problems to providing possible solutions.

Ideation will help you

1. Ask the right questions + innovate.
2. Discover unexpected areas of innovation.
3. Create volume and variety in your innovative options.
4. Bring together perspectives + strengths of team members.
5. Get obvious solutions out of your heads + drive your team beyond them

How to Brainstorm Solutions through Ideation Workshops

For the ideation stage, you will want a sample of voices from across your organization. To begin, schedule a workshop with your strategy team. Ideally, you will want the location to be off-site to limit distractions. If you are away from the office, those interruptions are less likely to happen. Participants should also turn off their phones and leave their laptops at home as these devices can also be distracting.

How to Facilitate Successful Ideation

Successful ideation requires good facilitation. Facilitation starts before the ideation session even begins and involves a lot more than acting as a host or note-taker. Here are three tips for facilitating successful ideation:

1. Never facilitate alone.
2. Learn who your audience is and tailor the workshop to them.
3. Break your ideation sessions up into multiple activities.

Types of Ideation Activities

Brainstorm:

1. Start with your problem statements and the "How Might We" (HMW) questions you wrote in the previous stage.
2. Focus on one problem at a time.
3. Write down, draw or act out ideas.
4. Build on other ideas (1+1=3).
5. Brainstorming sessions can be as short as 15 minutes or as long as two hours. Choose a time limit that works best for your organization.

Brain dump

1. Start with your problem statements and the HMW questions you wrote in the previous stage.
2. Ask participants to write their ideas on sticky notes. This should be done individually and silently.
3. Limit writing time to 10-15 minutes.
4. Once completed, have each participant present their ideas to the group and stick them on the wall. Hold any discussion until all participants are done.
5. Group duplicates together.
6. Vote on which ideas to develop further. Use the voting methods below.

Brain writing

1. Start with your problem statements and the HMW questions you wrote in the previous stage.

2. Ask participants to jot down their ideas on index cards.
3. Limit writing time to 3-5 minutes.
4. After time is up, have participants pass their cards to the next person and expand upon the idea.
5. Participants should pass ideas 3-10 times. This should be done silently.
6. Once completed, have each participant present their cards.
7. Use the voting methods below to select which ideas to expand upon.

Strategies for Innovative Thinking

There is a very simple law in operation here, the first law of creativity the quality and uniqueness of stimulus in has a direct impact on the quality and uniqueness of ideas out.

When you want to challenge your usual way of thinking so as to come up with new ideas.

Methods of Innovative Thinking

- Re-Expression
- Revolution
- Random Links

The concept behind re-expression is that people who work in the same industry or come from similar backgrounds frequently use the same words to explain situations and thoughts. When we repeat the same words, they become associated with a variety of associations or schemes. For example, let's say that your task is to come up with fresh new ideas for improving a fitness app and make it stand out against its competitors. If you think about the word 'fitness', you probably have many associations about what fitness is and how you achieve it, especially if you have done your background research properly. Your associations are probably not that different from your competitors' who, not unpredictably, are also trying to come up with innovative solutions for a fitness app. You might think about fitness centers, fitness trackers, training programmes, and so on.When we think about fitness, people from similar backgrounds and businesses tend to have similar associations.

Re-Expression

Re-expression is a method to help you think about the challenge in a new and different way that is also different from what your competitors are thinking. This is where the fun really starts. Have you ever heard the phrase "can't see the wood for the trees". It means being too close to something to be able to notice aspects of it or even its true identity. So, getting distance on your target subject is vital. The book suggests three ways you can do this:

1. **Re-express in different words**. Come up with as many related words or metaphors as possible for the issue (product or service) you are seeking to innovate. The word does not have to mean exactly the same thing; it can be something related (an example for 'fitness' could be 'endurance', 'play', 'health', 'repair', 'robustness', and 'strength'). Then think about what associations you have for each of the new words. Do some of them inspire your original issue? How? Why?
2. **Re-express in different senses.** Another method for bringing about new ways to think about an issue is to use different senses. You could draw it, act it out, or build it. Expressing it in different senses will allow you to see connections that were not previously obvious. Again, you should figure out how they inspire your original issue. As you're doing this, ask yourself why each connection links to the issue. For instance, if you've drawn a tree with powerful branches and a strong, intricate root system, you may discover that while fitness may mean 'motion' to many, there's no reason it can't involve 'nutrition' and 'settled in good soil (i.e., lifestyle habits)', too.
3. **Re-express from another perspective**. Try to think about how someone else, with different sensibilities from yours, would think about the issue. E.g., how would a 4-year-old think about fitness? Probably more like play and fun than exercise and how would a person in a wheelchair think about fitness? Challenging? With envy? With nostalgia? This method is in line with 'Extreme Personas'. Your assumptions do not need to be an accurate reflection of how the other person would think. The main thing is that imagining another perspective will allow you to see the issue differently, and that's the key to the enterprise here.

Build and Test Your Idea

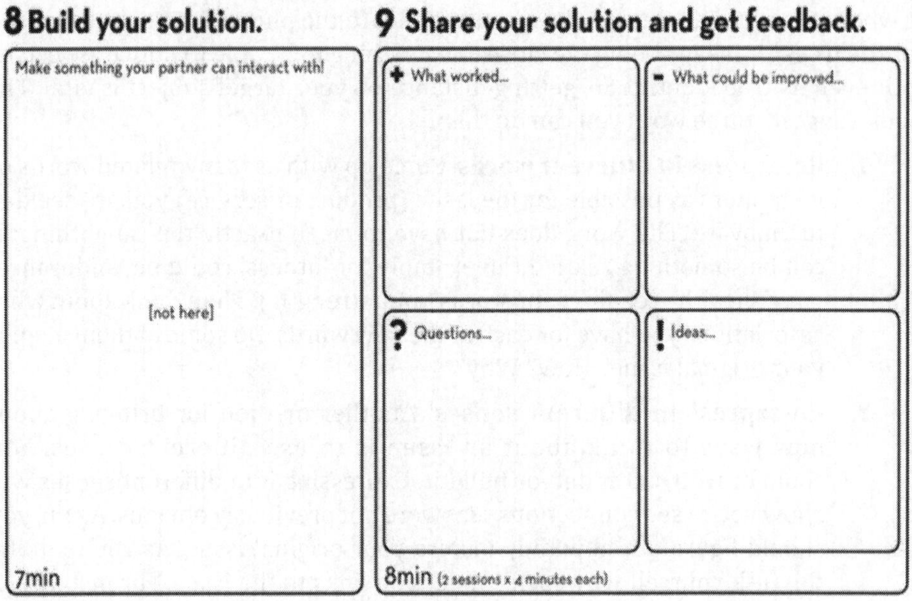

Fig 2: Sample Ideation

Chapter-4

Business Plan

A business plan is a fundamental document that any startup business needs to have in place prior to beginning operations. Banks and venture capital firms indeed often make writing a viable business plan a prerequisite before considering providing capital to new businesses.

Operating without a business plan is not usually a good idea. In fact, very few companies are able to last very long without one. There are definitely more benefits to creating and sticking to a good business plan—including being able to think through ideas without putting too much money into them and, ultimately, losing in the end.

A good business plan should outline all the projected costs and possible pitfalls of each decision a company makes. Business plans, even among competitors in the same industry, are rarely identical. But they all tend to have the same basic elements, including an executive summary of the business and a detailed description of the business, its services, and its products. It also states how the business intends to achieve its goals

Elements of a Business Plan

The length of the business plan varies greatly from business-to-business. All of the information should fit into a 15- to 20-page document. If there are crucial elements of the business plan that take up a lot of space such as applications for patents they should be referenced in the main plan and included as appendices.

As mentioned above, no two business plans are the same. But they all have the same elements. Below are some of the common and key parts of a business plan.

- **Executive summary:** This section outlines the company and includes the mission statement along with any information about the company's leadership, employees, operations, and location.
- **Products and services:** Here, the company can outline the products and services it will offer, and may also include pricing, product lifespan, and benefits to the consumer. Other factors that may go into this section include production and manufacturing processes, any patents the company may have, as well as proprietary technology. Any information about research and development (R&D) can also be included here.
- **Market analysis:** A firm needs a good handle on the industry as well as its target market. It will outline who the competition is and how it factors in the industry, along with its strengths and weaknesses. It will also describe the expected consumer demand for what the business is selling and how easy or difficult it may be to grab market share from incumbents.
- **Marketing strategy:** This area describes how the company will attract and keep its customer base and how it intends to reach the consumer. This means a clear distribution channel must be outlined. It will also spell out advertising and marketing campaign plans and through what types of media those campaigns will exist on.
- **Financial planning:** In order to attract the party reading the business plan, the company should include its financial planning and future projections. Financial statements, balance sheets, and other financial information may be included for already-established businesses. New businesses will instead include targets and estimates for the first few years of the business and any potential investors.
- **Budget:** Any good company needs to have a budget in place. This includes costs related to staffing, development, manufacturing, marketing, and any other expenses related to the business

Agribusiness Business Plan

Agriculture is one sector that thrives regardless of the country's economic situation. Food is, after all, one of humanity's most basic needs for survival. As a result, any country's agricultural industry is regarded as its most important. This is especially true in a country like India, which is renowned as an agricultural country because agriculture is the backbone of the Indian economy.

Scope of Business Plans in Agriculture:

- More than 60% to 70% of the Indian population is dependent on agriculture and its allied activities for their livelihood
- The agricultural sector provides employment to more than 52% of total labour of the country

- The contribution of agriculture to India's total GDP (Gross Domestic Product) is between 14% to 15%.

Agriculture can potentially be a feasible business idea in India.

Developing a Business Idea

The first stage in making a business concept into reality is to develop an idea into a viable product or service. Basic research can assist advance an idea, but this toolbox will help you turn it into a business, whether you want to build a manufacturing plant, a restaurant, an agricultural enterprise, a retail store, or a high-tech firm.

Starting or Acquiring

Whether you are starting a business from the ground up or acquiring a business that is already established, there are numerous things you should consider when making this preliminary decision:

Starting a Business

Pros

Starting a business allows you to develop an idea that you have an invested interest in, into a viable business. This may also attribute to accumulated knowledge of the industry and target market.

Cons

Starting your own business involves a multitude of challenges that established businesses have already dealt with, including building a customer base, marketing, hiring employees, establishing cash flow, finding space and creating a reputation or track record.

Acquiring a Business

Pros

The operation has likely been proven to generate cash flow, an already established customer base, conducted financial projections, easier financing and the possibility of receiving valuable legal rights like patents and copyrights.

Cons

Acquiring a business is often more costly and can carry obsolete inventory, uncooperative employees and outdated distribution methods along with it.

When acquiring a business, it is helpful to conduct a preliminary analysis that includes questions such as:

- Why is the business for sale?
- What is the perception of the industry?

- What is the outlook for future?
- Does or will the product line need to evolve over time?
- Are the raw materials in abundant supply?
- Will the market share remain in order to stay profitable?

Chapter-5

Business Model Canvas

The Business Model Canvas (BMC) is a strategic management tool to quickly and easily define and communicate a business idea or concept. It was proposed by Alexander Osterwalder based on his earlier book: Business Model Ontology. It outlines nine segments which form the building blocks for the business model in a nice one-page canvas.

It is a one-page document which works through the fundamental elements of a business or product, structuring an idea in a coherent way.

The Business Model Canvas reflects systematically on your business model, so you can focus on your business model segment by segment. This also means you can start with a brain dump, filling out the segments the spring to your mind first and then work on the empty segments to close the gaps. The following list with questions will help you brainstorm and compare several variations and ideas for your next business model innovation.

The Business Model Canvas (BMC) gives the structure of a business plan without the overhead.

The Canvas has nine elements:

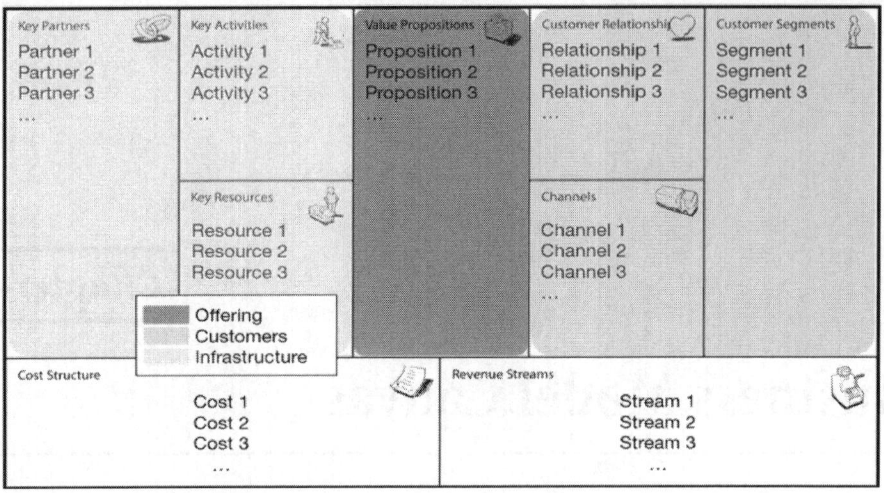

Together these elements provide a pretty coherent view of a business' key drivers–

- **Customer Segments:** Who are the customers? What do they think? See? Feel? Do?
- **Value Propositions:** What's compelling about the proposition? Why do customers buy, use?
- **Channels:** How are these propositions promoted, sold and delivered? Why? Is it working?
- **Customer Relationships:** How do you interact with the customer through their 'journey'?
- **Revenue Streams:** How does the business earn revenue from the value propositions?
- **Key Activities:** What uniquely strategic things does the business do to deliver its proposition?
- **Key Resources:** What unique strategic assets must the business have to compete?
- **Key Partnerships:** What can the company not do so it can focus on its Key Activities?
- **Cost Structure:** What are the business' major cost drivers? How are they linked to revenue?

The Canvas is popular with entrepreneurs and intrapreneurs for business model innovation. As per my personal experience and expertise it delivers three things:

1. **Focus:** Stripping away the 40+ pages of 'stuff' in a traditional business plan, I've seen users of the BMC improve their clarify and focus on what's driving the business (and what's non-core and getting in the way).
2. **Flexibility:** It's a lot easier to tweak the model and try things (from a planning perspective) with something that's sitting on a single page.
3. **Transparency:** Your team will have a much easier time understanding your business model and be much more likely to buy in to your vision when it's laid out on a single page.

Need to use Business Plan Canvas for Your Idea

1. To quickly draw a picture of what the idea entails.
2. It allows us to get an understanding of your business and to go through the process of making connections between what your idea is and how to make it into a business.
3. It looks at what kinds of customer decisions influence the use of your systems.
4. It allows everyone to get a clear idea of what the business will likely be.

How to use it

- The Value Proposition is foundational to any business/product.
- It is the fundamental concept of the exchange of value between your business and your customer/clients.
- Generally, value is exchanged from a customer for money when a problem is solved or a pain is relieved for them by your business.

Questions to ask when defining your business/product:

- What is the problem I am solving?
- Why would someone want to have this problem solved?
- What is the underlying motivator for this problem?

Tips:

A good way to approach this for users/ customers is by looking at your customer segments and figuring outwhere your product/ service solves the problem for your customer, based on Maslow's Hierarchy of Needs.

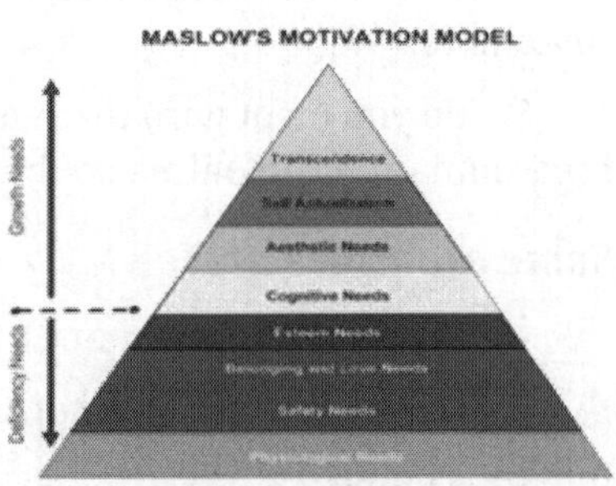

If you are selling your product or service to another business, you are a key partner in them achieving their Value Proposition for their customers.

It is important to have context around the goals the company is trying to achieve for their Customer Segments and where your business/product/service fits in the value chain.

Understanding the Steps in Business Model Canvas

1. Customer Segments

Customer Segmenting is the practice of dividing a customer base into groups of individuals that are similar in specific ways, such as age, gender, interests and spending habits.

Things to consider when determining your Customer Segments:

- Who are we solving the problem for?
- Who are the people that will value my value proposition?
- Are they another business?
- If so, what are the characteristics of those businesses?
- Or, are they other people?
- Does my value proposition appeal to men/women or both?
- Does it appeal to young adults aged 20 to 30 or teenagers?
- What are the characteristics of the people who are looking for my value proposition?

Another thing to gauge and understand is your market size, and how many people there are in the Customer Segment. This will help you understand your market from a micro and macro perspective.

A great place to start understanding your customer is to create customer personas for each of your Customer Segments.

2. Customer Relationships

We now know our Value Proposition and have developed Personas to better understand our Customer Segments or 'customers', but what is the relationship we have with our customers?

Customer Relationships is defined as how a business interacts with its customers.

So, do you meet with them in person? Or over the phone? Or is your business predominantly run online so the relationship will be online too?

Some examples are:

- In person (one-to-one)
- Third party contractors
- Online

- Events (one-to-many)
- Phone

This helps clarify the points of engagement between you and your customer and the modes used to relate to your customers. This will also help you start to define your operations as a business and also help you identify opportunities for automation.

3. Channels

Channels are defined as the avenues through which your customer comes into contact with your business and becomes part of your sales cycle.

This is generally covered under the marketing plan for your business.

Questions to ask when identifying the channels to reach your customers are:

- How are we going to tell our customer segment about our value proposition?
- Where are our customers?
- Are they on social media?
- Are they driving their car and listening to the radio?
- Are they at an event or conference?
- Do they watch TV at 7pm on a Friday night?

Examples of channels:

- Social media
- Public speaking
- Electronic mail (email marketing)
- Networking
- SEM (Search Engine Marketing)
- SEO (Search Engine Optimisation)
- Engineering as marketing
- Viral marketing
- Targeting blogs
- Sales and promotions for commissions
- Affiliates
- Existing platforms
- PR
- Unconventional PR
- Social advertising

- Trade shows
- Content marketing
- Community building
- Offline advertising (billboards, TV, radio)

Understanding how to reach your customers is so crucial to your business.

4. Key Activities

The Key Activities of your business/product are the actions that your business undertakes to achieve the value proposition for your customers.

Questions to ask:

- What activities does the business undertake in achieving the value proposition for the customer?
- What is the resource used?
- Time?
- Expertise?
- Distribution of product?
- Technical development?
- Strategy?
- Offer resources (human/physical)?
- What actions does it take you and/or your staff to achieve value exchange?

Examples:

- Consulting
- Designing
- Web development
- Baking
- Driving
- Shovelling

5. Key Resources

Next you should think about what practical resources are needed to achieve the key activities of the business. The Key means the resources your business requires to do business.

These resources are what is needed practically to undertake the action/ activities of your business:

- Office space
- Computers

- Hosting
- People (staff)
- Internet connection
- Car
- Bike
- Oven
- Electricity
- Car Parts

6. Key Partners

Key Partners are a list of other external companies/suppliers/parties you may need to achieve your key activities and deliver value to the customer. This moves into the realm of 'if my business cannot achieve the value proposition alone, who else do I need to rely on to do it?'.

An example of this is 'if I sell groceries to customers, I may need a local baker to supply fresh bread to my store'.

They are a key partner to achieve the value my business promises to the customer.

7. Cost Structures

Business cost structure is defined as the monetary cost of operating as a business.

- How much does it cost to achieve my businesses key activities?
- What are the cost of my key resources and key partnerships?
- How much does it cost to achieve the value proposition for my customers/ users?
- Are there additional costs to running a business?
- Legal?
- Insurance?
- What is the cost of my business?
- It is important also to place a monetary value on your time as a cost.
- How much would it cost you to hire you?
- What is the opportunity cost of running your business?

8. Revenue Streams

Revenue Streams are defined as the way by which your business converts your Value Proposition or solution to the customer's problem into financial gain.It is also important to understand pricing of business accordingly to pain of purchase in exchange for the pain of solving the problem for your customer.

How to gain revenue?

There are many different revenue models here:

- Pay per product (pay per view)
- Fee for service
- Fixed rate
- Subscription
- Dividends
- Referral feeds
- Freemium
- Equity gain

Put Your Idea to Work

According to extensive research, the business model canvas approach is the most beneficial way to start creating your idea. The Business Model Canvas is an excellent tool for analyzing and determining the viability of your value proposition throughout the early stages of the startup process. This is in contrast to diving directly into a full, traditional business plan, which can take hours to complete and often leaves the entrepreneur stuck, unsure of where to go next. This concept is taught all over the world and is an appealing option for anyone looking for a more cost-effective startup solution. This strategy is also recommended since it pushes the entrepreneur to consider how many aspects of their firm will interact to provide value to their clients. The model is shown below:

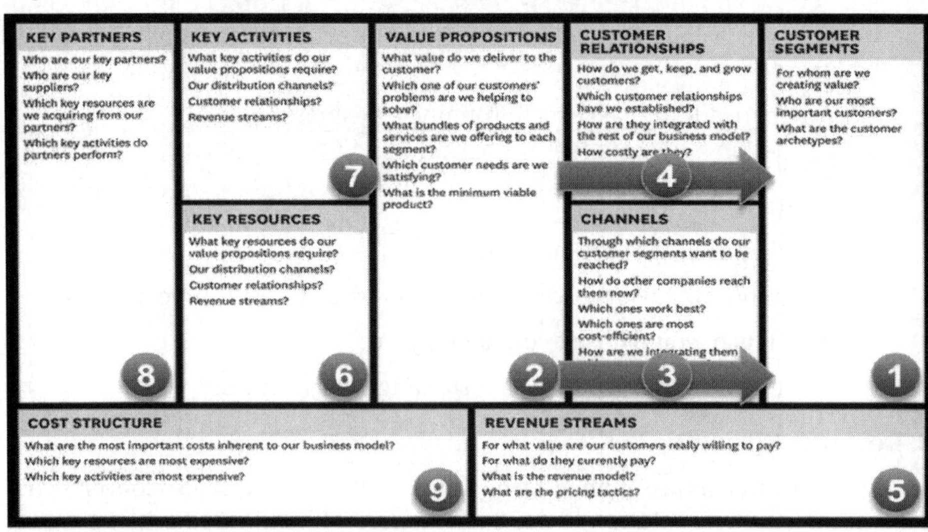

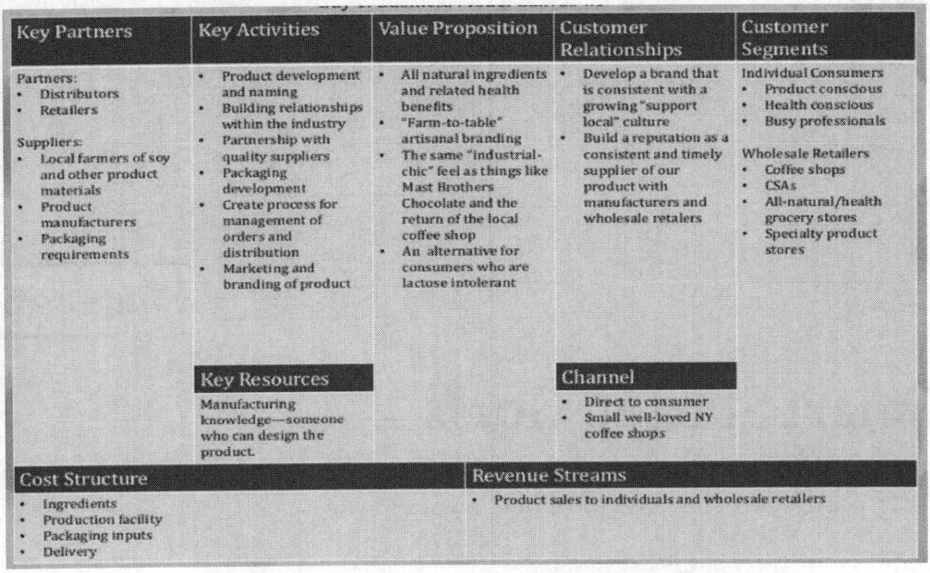

Business Model Canvas Information and Template

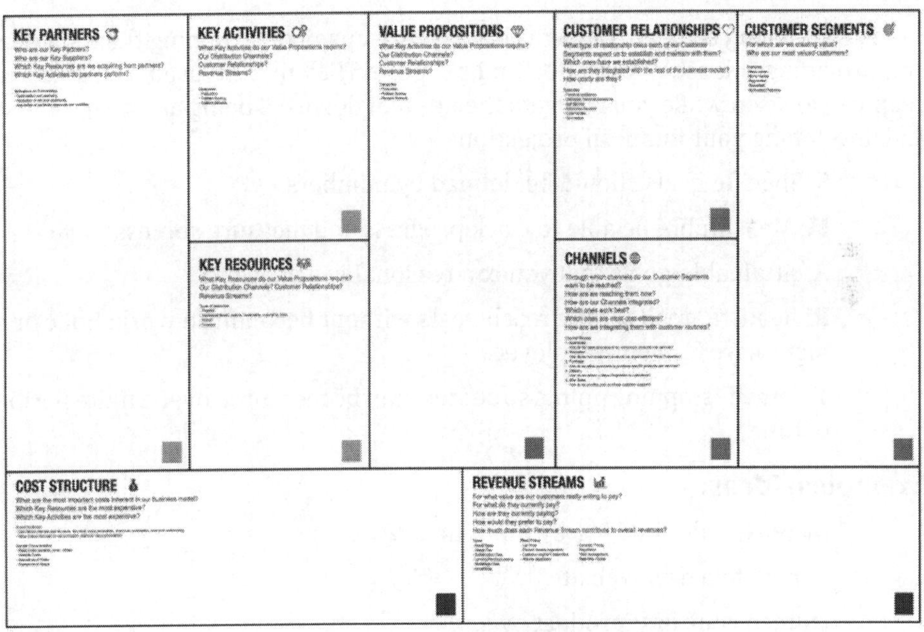

Business Model Canvas Template

Chapter-6

Smart Startup Goals

SMART Goals

A well-balanced set of objectives aids Start-ups in determining their success and providing a sense of direction and purpose. They may be used as a guide to keep you focused while you navigate the ups and downs of being an entrepreneur and developing your financial projections.

- **S**: Specific-goals should be defined by numbers
- **M**: Measurable-be able to track progress and measure success
- **A**: Attainable-goals that you can reasonably achieve
- **R**: Realistic-realistically reach goals without becoming a workaholic or a nightmare for your employees
- **T**: Timed- stopping points so success can be measured over a finite period of time

Examples/Ideas:

- Increase Sales by 20% by the year 2023
- Complete a new website by....
- Launch your first product by....

Ups and downs of being an entrepreneur

Pros	*Cons*
Can make a lot of money: the richest people in Britain are entrepreneurs of one business type or another.	**Risk:** The financial risk in starting up a company does not guarantee a fixed flow of income foryourself or your loved ones at first.

Pros	Cons
Independence: Many people don't like to be answerable a boss and prefer being in control of their working lives. They are leaders as opposed to followers.	**Time:** running a business may be heavily demanding on your time.
Personal Satisfaction: entrepreneurs can be extremely fulfilled with knowing that they have made their own money and can even increase self-esteem.	**Slow Rate of Return:** In the first years most businesses do not make a massive profit. Time needs to be invested for a company to grow.
Family: A lot of people become entrepreneurs so that they can balance their family life and can take time off to help with the family.	**Not being taken seriously:** can affect confidence and ability to perform.
Social satisfaction: through social entrepreneurship, there is a lot of community good, active citizenship and wellbeing. For instance, tackling a local problem by setting up a group/service/ business that addresses that problem.	**Economic climate:** may not be suitable to new ideas. What is happening at national and even global level can affect businesses negatively.
Control: Being in charge of how time is spent and other areas of work management lead to greater autonomy and control over life.	**Isolation:** many people starting their own business may feel lonely of not properly supported. It is a different work culture and requires a great deal of self-motivation.

Access to Capital

When you're looking for startup money, it's crucial to evaluate what the lender is looking for during your interview. The 6 C's of credit are basically the characteristics that lenders value in their borrowers and that influence the borrower's propensity to repay and the lender's level of acquired risk.

They are listed below:

- **Character:** A subjective analysis that determines whether or not you have the reputation to repay or not. Lenders use your personal credit report for this.
- **Capacity:** Measures the borrower's ability to pay the loan by assessing their income to debt ratio.
- **Capital:** What amount of capital does the borrower have to apply to the loan? For example, if the borrower has a lot of capital to put on as a down payment of a loan, a bank recognizes this as a way to decrease the chance of default.

- **Collateral:** Gives the lender the assurance that if the borrower defaults on the loan, the lender can seize the collateral and sell to make up for the loss on the default.
- Conditions: Refers to what the borrower intends to use the loan for. Banks want to see it being used for a specific purpose.
- Cash Flow: Does the prospective borrower have the cash flow to repay the loan—considering the numerous other variable and fixed expenses they will incur? At the very least, does the borrower have financial projections that tell the borrower that they will have the cash flow to repay?

1. **Family and Friends:** Those who are supportive and willing to provide capital should be treated as creditors. This means that the financing agreement should include a payment in plan and even interest over a time frame.
2. **Equity financing:** A method of raising capital by selling company ownership to investors.

   Equity Financing and Dilution Calculator (http://www.ownyourventure.com/equitySim.html): This tool helps entrepreneurs understand the impact of raising money for an early-stage venture.

3. **Angel investors:** Involves acquiring capital in exchange for equity and/or convertible debt. Angel investors are willing to help you all across the state.
4. **Crowd funding:** This method involves raising small amounts of capital from a large number of investors. This has become an increasingly popular method of raising capital.

Chapter-7

Startup Marketing Strategies

Marketing is the process of getting the right goods or services or ideas to the right people at the right place, time, and price, using the right promotion techniques and utilizing the appropriate people to provide the customer service associated with those goods, services, or ideas. Marketing refers to activities a company undertakes to promote the buying or selling of a product or service. Marketing includes advertising, selling, and delivering products to consumers or other businesses. Some marketing is done by affiliates on behalf of a company.

The marketing concept is the use of marketing data to focus on the needs and wants of customers in order to develop marketing strategies that not only satisfy the needs of the customers but also the accomplish the goals of the organization. An organization uses the marketing concept when it identifies the buyer's needs and then produces the goods, services, or ideas that will satisfy them. The marketing concept is oriented toward pleasing customers (be those customers organizations or consumers) by offering value. Specifically, the marketing concept involves the following:

- Focusing on the needs and wants of the customers so the organization can distinguish its product(s) from competitors' offerings. Products can be goods, services, or ideas.
- Integrating all of the organization's activities, including production and promotion, to satisfy these wants and needs
- Achieving long-term goals for the organization by satisfying customer wants and needs legally and responsibly

Components of a Market:

For a market to exist, certain conditions must be satisfied. These conditions should be both necessary and sufficient. They may also be termed as the components of a market.

1. The existence of a good or commodity for transactions(physical existence is, however, not necessary);
2. The existence of buyers and sellers;
3. Business relationship or intercourse between buyers and sellers.
4. Demarcation of area such as place, region, country or the whole world. The existence of perfect competition or a uniform price is not necessary.

Dimensions of a Market:

There are various dimensions of any specified market.

These dimensions are:

- Location
- Area or coverage
- Time span
- Volume of transactions
- Nature of transactions
- Number of commodities
- Degree of competition
- Nature of commodities
- Stage of marketing
- Extent of public intervention
- Type of population served
- Accrual of marketing margins

Marketing Strategy for Startups

A marketing strategy refers to a business's overall game plan for reaching prospective consumers and turning them into customers of their products or services. A marketing strategy contains the company's value proposition, key brand messaging, data on target customer demographics, and other high-level elements.

A thorough marketing strategy covers "the Seven Ps" of marketing.

7 P's of Marketing:

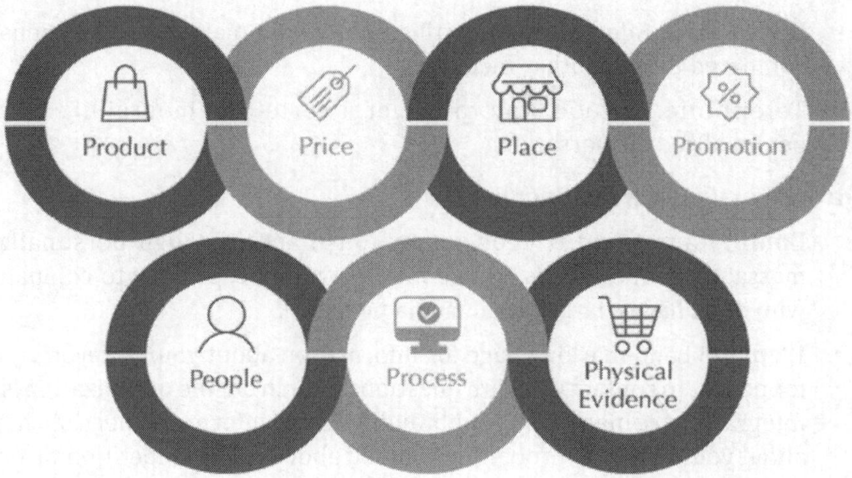

- **Product:** Describe product or service offered.
- **Price:** Price strategies that will help you reach your target profit margin.
- **Product:** Product strategies
- **Place:** Indicate where your business will sell its products or services and how they will get to consumers.
- **Promotion:** Methods used to communicate the features and benefits of your product or services to your target customers.
- **People:** Decide sales strategy and the people who will provide sales

Earning Brand image for your Startup

The first stage in marketing is to develop your brand image. You aim to sell to your customers by encapsulating all aspects of your brand in your marketing message. The Marketing Plan aids in the development of a strategy for marketing and communicating your product or service. You should also describe the communication media you want to utilize and how you intend to pay for them. Below are some resources to help brand and advertise your business:

Media

Traditional media include:

- **Print:** Newspapers/Magazines are still widely used specifically in rural areas.
- **Radio:** Most stations are a specific genre in an area and can help you target a certain demographic.

- **Television:** Expensive and difficult to prepare but effective in today's market.
- **Direct Mail:** Effective for directing to a certain market and inexpensive compared to some other mediums.
- **Telephone:** A good way to target a certain demographic with a personalized conversation.

Non-traditional media include:

- **Email:** Target and trigger ads to consumers through personalized messages/coupons. It is a huge marketing tool especially to companies who don't have a large social media presence.
- **Website:** Land marking page for information about your enterprise and for people to contact you with questions. Search engine optimization is an integral part of marketing your brand. This will inform northern Michigan about your brand by propelling yourself above your competition through internet searches.

Social Media:

- **Facebook:** Popular with middle aged adults for promoting businesses and connecting within their area.
- **Instagram:** Photo and video network for teens and adults in their 20's & 30's.
- **LinkedIn:** Site designed for professional networking, job searching, and connecting.
- **Twitter:** Real time news feed for networking professionals.
- **Pinterest:** Social network of images and boards of someone's lifestyle, great for showcasing products of your enterprise.

Chapter-8

Company Registration Process

Choosing the appropriate legal structure, filling out and obtaining the proper licensing, and learning about your county's permitting and zoning regulations are all unforeseen challenges when starting a business.

One of the first steps for any entrepreneur with a viable product/idea should be to adopt the desired legal structure. A sole proprietorship, partnership, conventional corporation, subchapter corporation, and limited liability company are the most common structures to choose from.

Private Limited Company

Start-ups and growing companies pick this popular business structure because it allows outside funding to be raised easily, limits the liabilities of its shareholders and enables them to offer employee stock options to attract top talent. As these entities must hold board meetings and file annual returns with the Ministry of Corporate Affairs (MCA), they tend to be viewed with more credibility than an LLP or General Partnership.

Features of Private Limited Company

For Businesses Raising Funding: Fast-growing businesses that will require funding from venture capitalists (VCs) need to register as private limited companies. This is because only private limited companies can make them shareholders and offer them a seat on the board of directors. LLPs would require investors to be partners and OPCs cannot accommodate additional shareholders. If you're raising funding, therefore, the points that follow scarcely matter; your decision is made

Limited Liability: Businesses often need to borrow money. In structures such as General Partnership, partners are personally liable for all the debt raised. If it cannot be repaid by the business, the partners would have to sell their personal possessions to do so. In a private limited company, only the amount invested in starting the business would be lost; the directors' personal property would be safe

Start-up Cost: A private limited company costs around Rs. 8000 to start at the very least, excluding professional fees. However, this will be higher in some states; in Kerala, Punjab and Madhya Pradesh, in particular, the fees are much higher. You also need some paid-up capital, which can be as little as Rs. 5000 to begin with. The annual compliance costs are around Rs. 13,000

Requires Greater Compliance: In exchange for the convenience of easily accommodating funding, the private limited company set-up needs to meet the demands of the Ministry of Corporate Affairs (MCA). These range from a statutory audit, annual filings with the Registrar of Companies (RoC), annual submission of IT returns, as well as quarterly board meetings, the filing of minutes of these meetings, and more. If your business isn't yet geared to meet these requirements, you may want to wait a while before you jump into registering a private limited company

Few Tax Advantages: The private limited company is assumed to have many tax advantages, but this is not actually the case. There are some industry-specific advantages, but taxes are to be paid at a flat rate of 30% on profits, the dividend distribution tax (DDT) applies, as does Minimum Alternate Tax (MAT). If you're looking for the structure with the lowest tax burden, the LLP does offer some better benefits.

Limited Liability Company

A relatively cheaper approach to incorporate as compared to a Private Limited Company and requires fewer compliances; its main improvement over General Partnership is that it limits the liabilities of its partners to their contributions to the business and offers each partner protection from negligence, misdeeds or incompetence of the other partners

Features of Limited Liability Company

Start-up Cost: Much cheaper than starting a private limited company, with government fees of Rs. 5000, no paid-up capital and low compliance costs

For Non-Scalable Businesses: If you're running a business that's unlikely to require equity funding, you may want to register an LLP as it combines several benefits of the private limited company and general partnership. It has limited liability, like a private limited company, and has a simpler structure, like a general partnership

Fewer Compliances: The MCA has made given some concessions to the LLP. For example, an audit needs to be performed only if your turnover is greater than Rs. 40 lakhs or paid-up capital is more than Rs. 25 lakhs. Furthermore, whereas all structural changes need to be communicated to the RoC in the case of private limited companies, the requirement is minimal for LLPs

Tax Advantages: Particularly if your business is earning over Rs. 1 crore in profits, the LLP offers tax benefits. The tax surcharge that applies on companies with profits over Rs 1 crore doesn't apply to LLPs, nor does Dividend Distribution Tax. Loans to partners are also not taxable as income

Number of Partners: There is no limit to the number of partners there may be in an LLP. If you're building a large advertising agency, for example, you need not worry about any cap on the number of partners.

General Partnership

A General Partnership is a business structure in which two or more individuals manage and operate a business in accordance with the terms and objectives set out in the Partnership Deed. This structure is thought to have lost its relevance since the introduction of the LLP because its partners have unlimited liability, which means they are personally liable for the debts of the business. However, low costs, ease of setting up and minimal compliance requirement make it a viable option for some, such as home businesses that are unlikely to take on any debt. Registration is optional in the case of General Partnerships.

Features of General Partnership

Unlimited Liability: On account of unlimited liability, the partners in the business are liable for all of its debts. This means that if, for whatever reason, a partner is unable to repay a bank loan or is liable to pay a fine, this can be recovered from his or her personal possessions. So, the bank, institution or supplier would have right to their jewellery, house or car. Furthermore, aside from ease of set-up and minimal compliance, the partnership offers no benefits over the LLP. If one opts to register it, which is optional, it may not even be cheaper. Therefore, unless one is running a very tiny business (let's say you offer a lunch pack service in your area and would like to set a profit ratio with your partner), you should not opt for a partnership

Easy to Start: If you choose not to register your partnership firm, all you need to get started is a partnership deed which you can have ready in just two to four working days. Even registration, for that matter, can be completed in a day once you have the appointment with the registrar. As compared with a private limited company or LLP, the procedure for starting-up is much simpler

Relatively Inexpensive: A General Partnership is cheaper to start than an LLP and even over the long-term, thanks to the minimal compliance requirements, is inexpensive. You would not need to hire an auditor. This is why, despite its shortcomings, home businesses may opt for it.

Sole Partnership

A sole proprietorship is a business that is owned and managed by a single person. You could have one up and running within 10 days, which makes it very popular among the unorganized sector, particularly small traders and merchants.

There is no such thing as registration; proprietorships are recognized by other registrations, such as a service or sales tax registration.

Features of Sole Proprietorship

Unlimited Liability: Just as a partnership, a sole proprietorship has no separate existence. Therefore, all debts can only be recovered from the sole proprietor. Therefore, the owner has unlimited liability with regard to all the debts. This should heavily discourage any risk-taking, which means that it's suited to only small businesses. If you plan on running a business that requires a loan or may end up paying penalties, fines or compensation, it's best you look into registering an OPC

Easy to Start: There is no separate registration procedure for proprietorships. All you need is a government registration relevant to your business. If you're selling goods online, a proprietor would only need a sales tax registration. Therefore, starting up as a sole proprietor is relatively easy.

One Person Company

The constitution of a One Person Company (OPC) was recently introduced as a strong improvement over sole proprietorship. It gives a single promoter full control over the company while limiting his/her liability to contributions to the business. This person will be the only director and shareholder (there is a nominee director, but with no power until the original director is incapable of entering into contract). Hence, there is no scope of raising equity funding or offering employee stock options.

Features of One Person Company

For Solo Entrepreneurs: A big improvement over the sole proprietorship firm, given that your liability is limited, the OPC is meant for solo entrepreneurs. However, do note that if it has revenues of over Rs. 2 crore and paid-up capital of over Rs. 50 lakh, it needs to be converted into a private limited company. Furthermore, given that there must be a nominee director (to enable perpetual existence of the OPC), you may as well consider starting a private limited company, which will also have flexibility of raising funding

High Compliance Requirements: While there are no board meetings, you will be required to conduct a statutory audit, submit annual and IT returns and comply with the various requirements of the MCA

Minimal Tax Advantages: The OPC, like the private limited company, has some industry-specific advantages. But taxes are to be paid at a flat rate of 30% on profits, the DDT applies, as does MAT. If you're looking for a structure with the lowest tax burden, the LLP does offer some better benefits

Start-up Costs: Nearly the same as a private limited company, with government fees, a little less than Rs. 7,000. However, this will change for different states, for example in Kerala, Punjab and Madhya Pradesh in particular, the fees is much higher.

Startup registration Process

Start-up / Company Registration

Different types of company which a start-up can register in India. A company is popularly referred to as a group of persons coming together with resources concerning capital, human resources, and skill for the common objective of making profits. In the old companies Act 1956 a company should have at least two persons as its member or shareholder.

Steps to Register your Start-up

Start-ups are becoming very popular in India. In order to develop the Indian economy and attract talented entrepreneurs, the Government of India, under the leadership of PM Narendra Modi, has started and promoted the Start-up India initiative to recognize and promote start-ups.

A start-up is a newly established business, usually small, started by 1 or a group of individuals. What differentiates it from other new businesses is that a start-up offers a new product or service that is not being given elsewhere in the same way. The keyword is innovation. The business either develops a new product/ service or redevelops a current product/service into something better.

Steps to Register Your Start-up with Start-up India

Step 1: Incorporate your Business

You must first incorporate your business as a Private Limited Company or a Partnership firm or a Limited Liability Partnership. You have to follow all the normal procedures for registration of any business like obtaining the Certificate of Incorporation/Partnership registration, PAN, and other required compliances.

Step 2: Register with Start-up India

Then the business must be registered as a start-up. The entire process is simple and online. All you need to do is log on to the Start-up India website and fill-up the form with details of your business. Next, enter the OTP which is sent to your e-mail and other details like, start-up as the type of user, name and stage of the start-up, etc. After entering these details, the Start-up India profile is created.

Once, your profile is created on the website, start-ups can apply for various acceleration, incubator/mentorship programmes and other challenges on the website along with getting access to resources like Learning and Development Program, Government Schemes, State Polices for Start-ups and pro-bono services.

Step 3: Get DPIIT Recognition

The next step after creating the profile on the Start-up India Website is to avail Department for Promotion of Industry and Internal Trade (DPIIT) Recognition. This recognition helps the start-ups to avail benefits like access to high-quality intellectual property services and resources, relaxation in public procurement norms, self-certification under labour and environment laws, easy winding of company, access to Fund of Funds, tax exemption for 3 consecutive years and tax exemption on investment above fair market value.

For getting DPIIT Recognition, click on the 'Get Recognised' button if you are a new user. If you are an existing user click on the 'Dashboard button' and then 'DPIIT Recognition'.

Step 4: Recognition Application

The 'Recognition Application Detail' page opens. On this page click on 'View Details' under the Registration Details section. Fill up the 'Startup Recognition Form' and click on 'Submit'.

Step 5: Documents for Registration

- Incorporation/Registration Certificate of your startup
- Details of the Directors
- Proof of concept like pitch deck/website link/video (in case of a validation/ early traction/scaling stage startup)
- Patent and trademark details (Optional)
- PAN Number

Step 6: Recognition Number

That's it! On applying you will immediately get a recognition number for your start-up. The certificate of recognition will be issued after the examination of all your documents which is usually done in 2 days after submitting the details online.

However, be careful while uploading the documents. If on subsequent verification, it is found to be obtained that the required document is not uploaded/

wrong document uploaded or a forged document has been uploaded then you shall be liable to a fine of 50% of your paid-up capital of the start-up with a minimum fine of Rs. 25,000.

Step 7: Other Areas

Patents, trademarks and/or design registration: If you need a patent for your innovation or a trademark for your business, you can easily approach any from the list of facilitators issued by the government. You will need to bear only the statutory fees thus getting an 80% reduction in fees.

Self-Certification Under Employment and Labour Laws: Start-ups can self-certify under labour laws and environment laws so that their compliance costs are reduced. Self-certification is provided to reduce regulatory burden thereby allowing them to focus on their core business. Start-ups are allowed to self-certify their compliances under six labour laws and three environment laws for a period of 3 to 5 years from the date of incorporation.

Units operating under 36 white category industries as published on the website of the Central Pollution Control Board do not require clearance under 3 environment-related Acts for 3 years.

Tax Exemption: Start-ups are exempted from income tax for 3 years. But to avail these benefits, they must be certified by the Inter-Ministerial Board (IMB). The Start-ups incorporated on or after 1st April 2016 can apply for the income tax exemption.

Documents Which have Been Waived Off

Start-up India has changed the procedure of registration since its inception. It has exempted most of the previous requirements now. Many documents which were required to be filed previously are waived off. The list of documents that are not required to be filed at the time of the registration are-

- Letter of Recommendations
- Letter of funding
- Sanction Letters
- Udyog Aadhar
- MSME Certificate
- GST Certificate

Chapter-9

Patenting Your Innovations

A patent is an exclusive right granted by the government to an inventor for a new invention for a certain amount of time for his or her disclosed invention. This exclusive monopoly right is only applicable within the territorial limits of a country of grant. Without the permission of the patent holder, no one else can make, use, manufacture, or promote the innovation. This permission is only available for a limited time.

Patent

A patent is a type of intellectual property for a limited amount of time that allows the owner of such IP specific legal rights against the unlawful selling, using, manufacturing, or sharing of any of his work or innovation. In other words, a patent is a licence that grants the innovator exclusive rights to his innovation, whether it is a technique or a product, for a set length of time. However, such an invention should either be a new way of doing something or a solution to a problem that currently exists. A patent is issued for 20 years, after which the invention is released into the public domain.

Types of Patents

1. Utility Patent

It is the most common type of a patent and covers all new inventions and significant modification and improved process or product, having some utility. Utility patent is also referred as 'patent for invention.' The patent protects the rights of the inventor over his invention by preventing any third party to use, sell, or make the invention in question, without proper authorization.

2. Design Patent

An exclusive right providing legal protection to the ornamental design of a useful product is granted under the design patent. Design of a printer, shape of spoons, etc. are some examples that can be granted design patent.

3. Plant Patent

As the name suggests, plant patent is the patent granted on plants. New plants that are produced through asexual modes of reproduction like cutting, vegetative propagation, etc. are eligible to be granted plant patent. However, plant patents are not granted in India.

Requirements of Patenting

- Novelty
- Non obviousness
- Marketability and Utility

It is important to remember

- A patent has a time limit (20 years)
- A patent has a geographical limit
- A patent is restricted to the claims made in the application

Patenting in India – Old and under the New Act

Item	*Old Act (1970)*	*New Act (2005)*
Human Life / Genes	Not patentable	Not patentable
Plants and animals	Not patentable	Not patentable
Microorganisms	Not patentable	Patentable
Non biological processes	Patentable	Patentable
Non biological products	Not patentable	Patentable
Biochemical and biotechnological process	Patentable	Patentable
Bio products	Not patentable	Patentable
Crop varieties	No protection	Protection under PVPFR
Animal varieties	Excluded	Excluded
Method of Agriculture and Horticulture	Not patentable	Products and process qualifying for novelty are patentable

Advantages of Patents

Patent gives the protection to the Patentee, enabling him to enjoy the right and to raise the capital for working his invention on a commercial scale. In case of not obtaining the patent, the product may be taken over by others to commercialise its benefits

Inventions can be patented if they satisfy three criteria:

1. Novelty
2. Non-obviousness
3. Marketability and Utility

Points to remember:

- In case, the Patentee is not able to work the invention commercially, he would be able to make a profitable use of his invention by selling his patent or by granting license to others, permitting the use of his invention
- Patent System helps for industrial growth by introducing new technologies
- The scientific knowledge contained in the patent specification helps as a "stepping stone" for further research
- Patents provide inventive and creative ideas for further R & D in the field.
- Patent acts as tradable industrial asset for the enterprise and, thus the strength of patent portfolio of the company is the indication of the good economic health of the company.
- After the term of patent is over, or patent is not kept in force, the patented invention is available to the public for free use.

Procedure for patent application

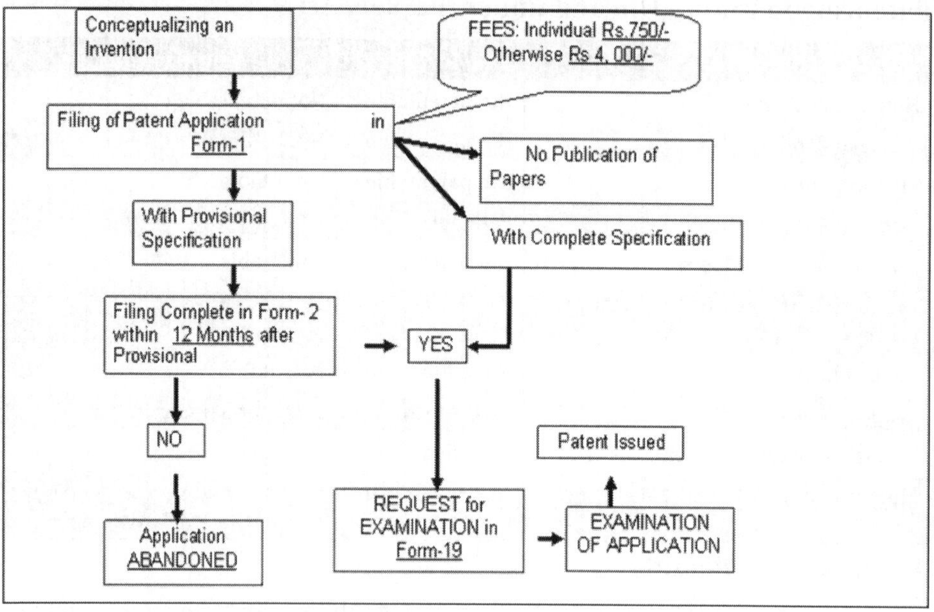

There are two types of patent documents usually known as patent specification, namely

Provisional Specification

Complete Specification

Provisional Specification

A provisional specification is usually filed to establish priority of the invention in case the disclosed invention is only at a conceptual stage and a delay is expected in submitting full and specific description of the invention. Although, a patent application accompanied with provisional specification does not confer any legal patent rights to the applicants, it is, however, a very important document to establish the earliest ownership of an invention. Complete specification must be submitted within 12 months of filing the provisional specification. This period can be extended by 3 months. It is not necessary to file an application with provisional specification before the complete specification. An application with complete specification can be filed right at the first instance.

Complete Specification

Submission of complete specification is necessary to obtain a patent The contents of a complete specification would include the following

1. Title of the invention.
2. Field to which the invention belongs
3. Background of the invention including prior art giving drawbacks of the known inventions & practices.
4. Complete description of the invention along with experimental results.
5. Drawings etc. essential for understanding the invention.
6. Claims, which are statements, related to the invention on which legal proprietorship is being sought. Therefore, the claims have to be drafted very carefully.

Cost of Filing Application in India

The Government fee for filing a patent application (complete/provisional) in India is Rs....... /- for individuals and Rs.......... /- for legal entities. An applicant is now required to make a request for examining the patent application within 48 months of filing of the application. An individual has to pay Rs....... /- as examination fee and Rs. Rs....... /- for legal entities. A sealing fee of Rs....... /- for individuals and Rs....... /- for legal entities has to be paid at the time of grant (sealing) of patent.

Publication of applications

All the applications for patents will not be open to the public for 18 months from the date of filing or date of priority whichever is earlier. After 18 months, every application will be published except the applications in which a secrecy direction is given under section 35, and the same will be notified in the Official Gazette. The publication includes the particulars of date of application, number of applications, name and address of the applicant and abstract.

Request for Examination

No application for a patent will be required to be examined unless the applicant or any other interested person makes a request in Form 19 within 48 months from the date of filing of the application for patent.

Examination and First examination Report

All the applications filed at the appropriate office shall be first subjected to formal screening and then to substantive technical examination by the examiner [when a request in Form 19 is made]. As the novelty, inventive step and industrial application are the criteria for the Grant of Patent, these aspects are thoroughly studied by the Examiner. Further, the examiner will conduct a search for novelty, using the databases available at the patent office after the subject matter of invention is classified according to Indian & International classification, and also various other sources.

Acceptance & Notification of Complete Specification

The controller of Patent shall accept the complete specification and inform the acceptance of complete specification to the applicant. Further, the acceptance of complete specification will be notified in the Gazette of India, Part III, section 2 which is published weekly, on every Saturday.

Opposition to the Grant of Patent

Any person interested can oppose the grant of patent within a period of four months, extendable by another month. The desire to oppose the grant should be put in an appropriate form with appropriate fee. An opponent shall file a Notice of Opposition giving details of the grounds on which he wishes to oppose the Grant of Patent. The notice of opposition should be followed by full written statement explaining the various ground of opposition. The opposition to the Grant of Patent coming before the controller is a bi-party proceedings wherein, the Controller, being a quasi-judicial authority, will decide the case based on the written statement and evidence placed by the opponent and, also the reply statement and evidence filed subsequently by the Applicant.

Grant of Patent

If the application is not opposed or the opposition is decided in favour of the applicant or is not refused otherwise, then the patent is granted and sealed upon request made by the applicant in the prescribed manner along with fees on payment of Sealing Fee within 6 months from the date of advertisement of acceptance of complete specification. The period is extendable by three months. Patent rights will accrue to the applicant only when the patent is sealed and the rights are granted to him.

In order to keep the patent in force, Renewal Fee is to be paid in the Patent Office annually.

How can a patent be obtained worldwide?

At present, no "world patents" or "international patents" exist. In general, an application for a patent must be filed, and a patent shall be granted and enforced, in each country in which you seek patent protection for your invention, in accordance with the law of that country. In some regions, a regional patent office, for example, the European Patent Office (EPO) and the African Regional Intellectual Property Organization (ARIPO), accepts regional patent applications, or grants patents, which have the same effect as applications filed, or patents granted, in the member States of that region. Further, any resident or national of a Contracting State of the Patent Cooperation Treaty (PCT) may file an international application under the PCT. A single international patent application has the same effect as national applications filed in each designated Contracting State of the PCT.

Patent offices in India

The Patent Office, under the Department of Industrial Policy & Promotion, Ministry of Commerce & Industry, performs the statutory duties in connection with the grant of patents for new inventions and registration of industrial designs. Patent Offices are located at Kolkata (Head Office), Mumbai, Chennai and Delhi to deal with the applications for patents originating within their respective territorial jurisdictions.

Head Office
Controller General of Patents and Designs and Trademarks,
Old CGO Building, 101, M K Road,
Mumbai - 400002 , India
Phone No.: 91-22-201 7368: 91-22-203 9050
Fax No.: 91-22-205 3372
http://www.patentoffice.nic.in

Regional Patent Office's Address:

Controller General of Patents, Designs & Trade Marks Bhoudhik Sampada Bhavan, Near Antop Hill Head Post Office, S.M. Road, Antop Hill, Mumbai-400037, Phones : 022-24123311, Fax : 022-24123322 Web site : www.ipindia.nic.in

Patent Offices

Address	*Contact Persons*
The Patent Office, Intellectual Property Office Building, CP-2 Sector V, Salt Lake City,Kolkata-700091, Phone : 23671945, 1946, 1987, FAX-033-2367-1988, Email:- kolkata-patent@nic.in	Deputy Controller of Patents & Designs, Assistant Controller of Patents & Designs

The Patent Office,Intellectual Property Office Building,G.S.T. Road, Guindy, Chennai-600032, Phone: 044-22502081- 84, FAX: 044-22502066, Email: chennai-patent@nic.in	Deputy Controller of Patents & Designs , Assistant Controller of Patents & Designs
The Patent Office,Intellectual Property Office Building,Plot No. 32, Sector 14, Dwarka, New Delhi-110075, Phone : 011-28031032. 28031039, 28031044, 28031053, FAX:011-28031583, 28031432 Email: delhi-patent@nic.in	Deputy Controller of Patents & Designs , Assistant Controller of Patents & Designs
Patent Office Boudhik Sampada Bhawan, S.M.Road, Near Antop Hill Post Office, Antop Hill, MumbaiI - 400 037. Phone : 24137701, 24141026, 24150381, 24148165, 24171457 FAX : 24130387 EMAIL: mumbai-patent@nic.in	Joint Controller of Patents & Designs, Assistant Controller of Patents & Designs
Designs Wing of the Patent Office	
The Patent Office(Designs Wing), Intellectual Property Office,CP-2 Sector V, Salt Lake City, Kolkata-700091 Phone No. 033-23671945-46, 23671987 FAX-033-23671988, Email:- kolkata-patent@nic.in Note : Application for Design can also be received by any of the Patent Office located at New Delhi, Chennai & Mumbai.	Assistant Controller of Patents & Designs

List of Forms

Form No.	*Section and rule*	*Title*
1.1.A	Sections 5(2), 7, 54, 135 and rule 39 Section 7(1A); rule 20(1)	Application for grant of a patent. Application for grant of patent on an application corresponding to an International application under PCT.
2.	Section 10; rule 13	Provisional/Complete Specification.
3.	Section 8 and rule 12.	Statement and undertaking.
4.	Sections 8(2), 9(1), 25(1), 28(4), 43(3), 53(3) and rules 12(4), 13(6), 24(5), 56(1), 73(3) or 130.	Request for extension for time.
5.	Section 10(6) and rule 13(6).	Declaration as inventorship.
6.	Sections 20(1), 20(4), 20(5) and rules 34(1), 35 or 36.	Claim or request regarding any change in applicant for patent.
7.	Section 25 and rule 55.	Notice of opposition to grant of a patent.
8.	Sections 28(2), 28(3) or 28(4) and rules 66, 67, 68.	Request or claim regarding mention of inventor as such in a patent.
9.	Section 43 and rule 73(1).	Request for sealing of a patent.
10.	Section 44 and rule 75.	Application for amendment of patent.
11.	Sections 51(1), 51(2) and rules 76, 77.	Application for direction of the controller.
12.	Section 52(2) and rule 79.	Request for grant of patent.
13.	Section 57 and rule 81(1).	Application for amendment of the application for patent/complete specification.

Form No.	*Section and rule*	*Title*
14.	Sections 57(4), 61(1), 63(3), 78(5) and 87(2) and rules 49(1), 52(3), 81(3)(b), 85(1), 87(2), 98(1), 101(3) or 124 and also section 87 (2) as modified by section 24C.	Notice of opposition to amendment/restoration/surrender of patent/grant of compulsory licence or revision of terms thereof or to a correction of clerical errors.
15.	Section 60 and rule 84.	Application for restoration of patents.
16.	Section 68 and rule 89.	Application for registration of a document.
17.	Sections 69(1) or 69(2) and rules 90(1) and 90(2)	Application for registration of title/interest in a patent or share in it or registration of any document purporting to affect proprietorship of the patent.
18.	Sections 84(1), 91 or 92(1) and rules 47, 96 and also sections 84 and 92 as modified by section 24C.	Application for compulsory licence.
19.	Section 11B and Rule 24(1).	Request for examination of application for patent.
20.	Section 85(1) and rules 47, 96, and also section 85(1) as modified by section 24C.	Application for revocation of a patent or exclusive marketing right.
21.	Section 88(4) and rules 51, 100 and also section 88(4) as modified by section 24C.	Application for revision of terms and conditions of licence.
22.	Section 94 and rule 102(1) and also section 94 as modified by section 24C.	Request for termination of compulsory licence.
23.	Rules 109 and 112.	Application for registration of Patent Agent.
24.	Section 130 (2) and rule 117.	Application for the restoration of the name in the register of Patent Agents.
25.	Sections 77(1)(f), 77(1)(g) and rules 130(1) or 130(2).	Application for review/setting aside controller's decision/order.
26.	Sections 127, 132 and rule 135.	Form of authorisation of a Patent Agent/or any person in a matter or proceeding under the Act.
27.	Section 24A and rule 40.	Application for grant of exclusive marketing rights.
28.	Rule 46.	Form for the grant of exclusive marketing rights.
29.	Section 146(2) and rule 131(1)	Statement regarding the working of the Patented invention.
30.	Section 39	Request for permission for making patent application outside India

Forms and Fees

Amount of fees (in rupees)

Form no.	*Natural Person*	*Legal Entity other than natural person(s).*
1	1000 - Multiple of 1000 in case of every multiple priority. - 100 (each sheet of specn. -in addition to 30) - 200 (for each claim in addition to 10)	4000 - Multiple of 4000 in case of every multiple priority. - 400 (each sheet of specn. -in addition to 30) - 800 (for each claim in addition to 10)
2	No fee - 100 (each sheet of specn. -in addition to 30) - 200 (for each claim in addition to 10)	No fee - 400 (each sheet of specn. -in addition to 30) - 800 (for each claim in addition to 10)
3	No fee	No fee
4	300 per month	1200 per month
5	No fee	No fee
6	500	2000
7	1500	6000
8	500	2000
9	2500	10000
10	1500	6000
11	1500	6000
12	1500	6000
13	500 1000 200	2000 4000 800
14	1500	6000
15	1500	6000
16	1000	4000
17	1500	6000
18	2500 3500	10000 14000
19	1500	6000
20	1500	6000
21	1500	6000
22	2000	No fee
23	1000	No fee
24	1000	4000
25	1000	4000

Form no.	*Natural Person*	*Legal Entity other than natural person(s).*
26	No fee	No fee
27	No fee	No fee

The forms can be downloaded from http://ipindia.nic.in/ipr/patent/patents.htm

Can an idea be patented?

Section 10 of the Patent Act provides for some essential specifications which must be provided in the application for patent. One of such specification is the preamble. In case of a provisional application, the preamble starts with 'the following specification describes the invention' and the preamble of the complete application reads as 'the following specification particularly describes the invention and the manner in which it is to be performed. if an idea has the potential to be performed and one has the method and process for its working, in this case, the idea has the possibility of changing into an invention and thereby can be applied for a patent through a provisional application.

For the reason of a mandate disclosure of best mode of performance of invention to be mentioned in the complete application, the idea can be described and filed for patent through provisional application and subsequently one can work a way out with a technique for its performance, within 12 months of filing provisional application and then the complete application may be filed. However, it is to be noted that on failure of filing complete application within the stipulated period, the provisional application may be rejected and be held invalid on the grounds of insufficient description of the invention or the method.

For an instance, a person has an idea of making a mobile application cyber threat warning system and has prowess to convert the thought into an invention. In such situation, the person may file a provisional application for grant of patent by describing his idea of the application and later within a period of 12 months may file a complete application also describing the mode and process of performance of such invention. Failure of submission of complete application may amount to invalidation of the application.

Chapter-10

Idea Pitch Deck For Funding

Pitching Your Idea to Investors

A pitch deck is usually a few slide presentation designed to give a short summary of company, business plan and startup vision. It also serves very different purposes, from trying to get a meeting with a new investor, to presenting in front of a stage, and each one of them should follow a different structure.

Coming up with good ideas is difficult, but convincing others to act on them is even more difficult. The task of bringing an idea to someone with the power to do something with it is known as a pitch.

Ideas demands change. Even if your idea is unquestionably fantastic, it will compel someone, somewhere to adjust their way of doing things. Because many people dislike and fear change, the features that you find appealing in your idea may be precisely what make it difficult for others to accept.

What should be in an investor pitch deck?

A number of venture capitalists, startup founders and evangelists have created different versions of what they consider required elements to successful pitching presentations. Most of them agree on the following investor deck outline:

- Introduction
- Team
- Problem
- Advantage
- Solution

- Product
- Traction
- Market
- Competition
- Business Model
- Investing
- Contact

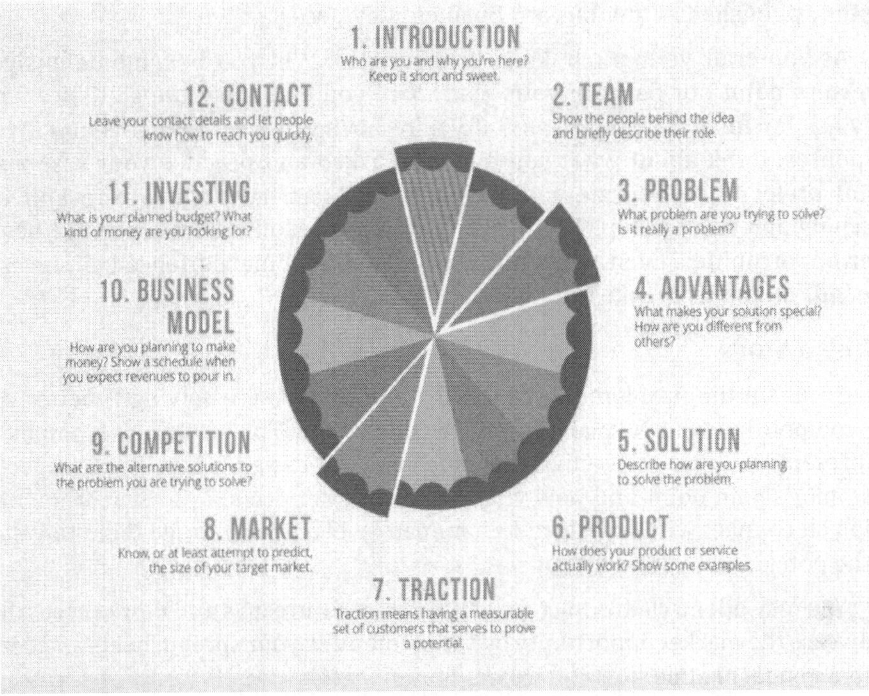

Fig1: Investment pitch-deck Sample

Important points for preparing your pitch.

1. Know Who You're Pitching

Regardless of their industry knowledge or the stage of their firm's investment, some entrepreneurs aim to get in front of every investor. You must keep in mind that accepting an investment entail more than just money; you are entering into a partnership. You must perform your due diligence and research potential investors before making your pitch.

- When researching, ask yourself:
- What industries do they invest in?

- What stage do they invest in?
- What's the investor's track record?

Given the benefits and high stakes, the more you know going into a pitch, the better.

2. Consider How You're Presenting Yourself, Not Simply Your Idea

While your ideas and abilities are vital, your personality is just as crucial. Venture capitalists' interest in a startup is "motivated less by judgements that the founder is competent than by views of character and trustworthiness," according to study published in the Harvard Business Review.

As you craft your pitch, keep this in mind. Will you become defensive if investors point out flaws in your plan? Will you embellish the statistics when they ask for financial projections? If you're having second thoughts about your responses, think about what you might be asked and practice your responses. Firms prefer to collaborate with founders they can trust who are receptive to coaching and mentorship. The majority of seasoned investors prioritize people over opportunities. Even when a team is young and inexperienced, an investor depends on them to make the right decisions."

3. Tell a Story

Focus on the problem you're solving and how you're solving it better than the competition when explaining your business idea. This could be accomplished by presenting a real-life scenario in which you discuss an existing or prospective customer's pain point and how your product or service alleviated it. This might help you connect with investors on a human level and encourage them to believe in the potential of your concept.

You may offer a clearer picture of your start-future up's and more successfully showcase the market opportunity by supplementing your spreadsheets and charts with an engaging tale.

4. Cover the Details

While it's critical to create the tone, you must also address the details. Define your value proposition succinctly in your pitch deck and include a distinctive slogan for investors to remember. From there, convey the opportunity and detail:

- The size of the market
- Your plan to attract and retain customers
- How you can create barriers to competition
- Whether you can run quick, inexpensive tests to determine product-market fit
- Your plan to monetize the business and make revenue
- The amount of capital investment required

In general, investors are trying to determine how entrepreneurs think about the opportunity rather than assess their presentation or prediction skills.

5. Show the Roadmap

Investors want to know how they'll get paid in the end, even if you're still in the early phases of your business. Highlight your exit strategy and the options available to close your pitch.

Investors prefer organizations with a variety of exit choices. They want to know how they can turn their investment into cash and give it back to their partners."

The most common exit strategies include:

- **Acquisition:** When one company buys most or all of another company's shares to gain control of it.
- **Merger:** When two existing companies are united into one new company
- **Initial Public Offering (IPO):** When a private company issues its first sale of stocks to the public and can start raising capital from public investors.

LANDING THE PITCH

Every investor prioritizes different data and information. Yet, if you start by choosing the right investor and then align their needs with your proposed market opportunity, value proposition, and exit strategy, you have a chance at landing the pitch.

Innovators can pitch their deck by preparing presentation of their ideas as per the given Sample Format.

FRENZY FARMS
FRENZY FARM LLP
Incorporated 9th July 2020
LLP Identification Number: AAS- 8297
Registered Address:
The Capital, 502, Danda Lakhond, Dehradun,
Uttarakhand,248013,India
Start-up India Certificate No. DIPP65915
CERTIFICATE OF RECOGNITION
Copy right © 2020 Frenzy Farms. All Rights Reserved
www.krishisandhi.com

FRENZY FARMS
KRISHISANDHI
OUR TEAM
SUBODH SHAH
Founder and Business Strategy & Development
IIT Roorkee graduatewith 18 plus year of exp in Manufacturing across globe
SHAILESH SINGH
HR Activities and Manpower Management
IIT graduate and 18 plus year work exp in various Industries
PARUL HANDE
Supply Chain Consultant
Supply chain expertwith 8 plus yr exp in multinational leading FMCG
C K VISHVKARMA
Consultant for IoT
IIT graduate and CEO of All things connected , partnering us in SMART farming consultation
MANDEEP PANDEY
Sales Head
18 year experience in in Uttarakhand/ NCR in Sales Domain
NIRMAL SHAH
Administration and HRBP for Supply chain network
15 Plus Exp. In service industry in field of Admin & Supply Chain activities
Copy right © 2020 Frenzy Farms. All Rights Reserved
www.krishisandhi.com

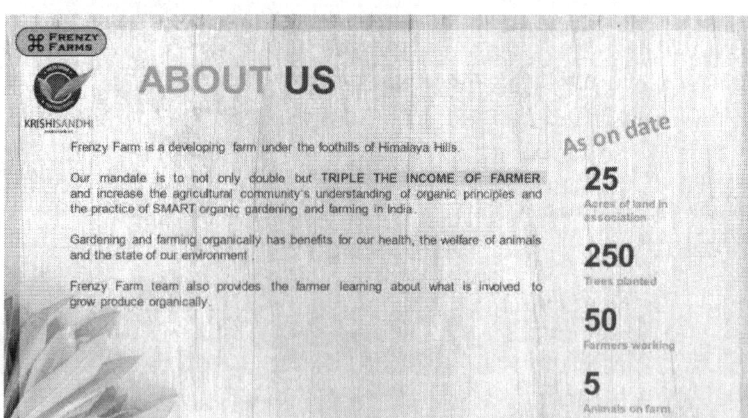
FRENZY FARMS
KRISHISANDHI
ABOUT US
Frenzy Farm is a developing farm under the foothills of Himalaya Hills.
Our mandate is to not only double but TRIPLE THE INCOME OF FARMER and increase the agricultural community's understanding of organic principles and the practice of SMART organic gardening and farming in India.
Gardening and farming organically has benefits for our health, the welfare of animals and the state of our environment.
Frenzy Farm team also provides the farmer learning about what is involved to grow produce organically.
As on date
25
Acres of land in association
250
Trees planted
50
Farmers working
5
Animals on farm
Copy right © 2020 Frenzy Farms. All Rights Reserved
www.krishisandhi.com

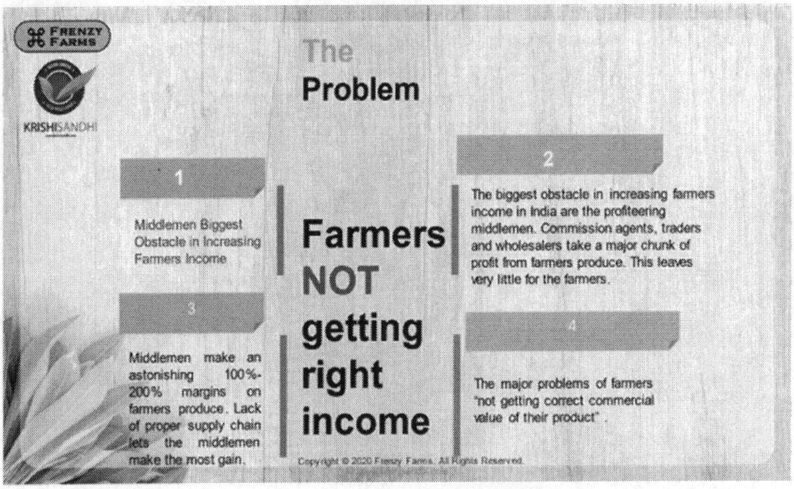
FRENZY FARMS
KRISHISANDHI
The Problem
1
Middlemen Biggest Obstacle in Increasing Farmers Income
2
The biggest obstacle in increasing farmers income in India are the profiteering middlemen. Commission agents, traders and wholesalers take a major chunk of profit from farmers produce. This leaves very little for the farmers.
3
Middlemen make an astonishing 100%-200% margins on farmers produce. Lack of proper supply chain lets the middlemen make the most gain.
4
The major problems of farmers "not getting correct commercial value of their product" .
Farmers NOT getting right income
Copyright © 2020 Frenzy Farms. All Rights Reserved.

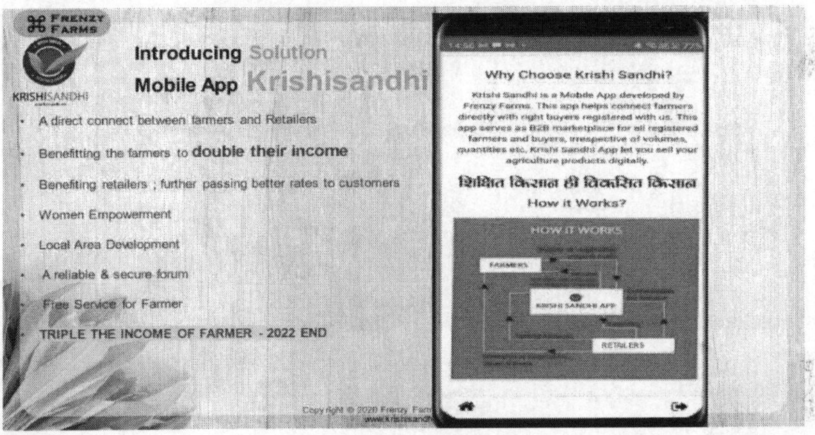
FRENZY FARMS
KRISHISANDHI
Introducing Solution
Mobile App Krishisandhi
A direct connect between farmers and Retailers
Benefitting the farmers to double their income
Benefiting retailers ; further passing better rates to customers
Women Empowerment
Local Area Development
A reliable & secure forum
Free Service for Farmer
TRIPLE THE INCOME OF FARMER - 2022 END
Why Choose Krishi Sandhi?
Krishi Sandhi is a Mobile App developed by Frenzy Farms. This app helps connect farmers directly with right buyers registered with us. This app serves as B2B marketplace for all registered farmers and buyers, irrespective of volumes, quantities etc. Krishi Sandhi App let you sell your agriculture products digitally.
शिक्षित किसान ही विकसित किसान
How it Works?
HOW IT WORKS
FARMERS
KRISHI SANDHI APP
RETAILERS

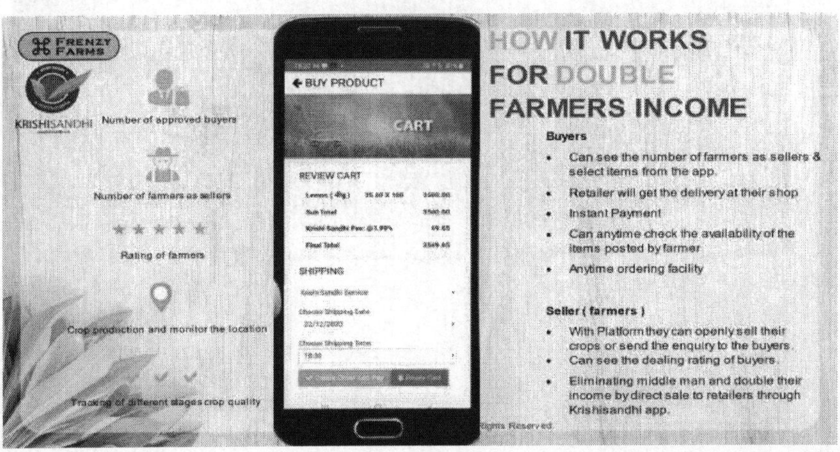
FRENZY FARMS
KRISHISANDHI
Number of approved buyers
Number of farmers as sellers
Rating of farmers
Crop production and monitor the location
Tracking of different stages crop quality
BUY PRODUCT
CART
REVIEW CART
SHIPPING
HOW IT WORKS FOR DOUBLE FARMERS INCOME
Buyers
Can see the number of farmers as sellers & select items from the app.
Retailer will get the delivery at their shop
Instant Payment
Can anytime check the availability of the items posted by farmer
Anytime ordering facility
Seller (farmers)
With Platform they can openly sell their crops or send the enquiry to the buyers.
Can see the dealing rating of buyers.
Eliminating middle man and double their income by direct sale to retailers through Krishisandhi app.
Rights Reserved.

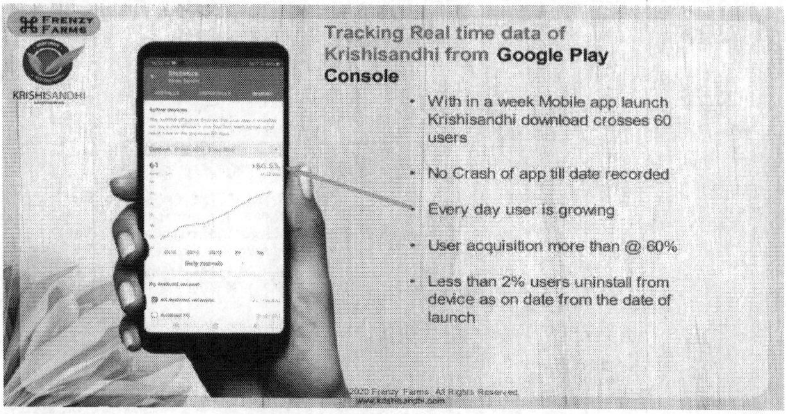
FRENZY FARMS
KRISHISANDHI
Tracking Real time data of Krishisandhi from Google Play Console
With in a week Mobile app launch Krishisandhi download crosses 60 users
No Crash of app till date recorded
Every day user is growing
User acquisition more than @ 60%
Less than 2% users uninstall from device as on date from the date of launch
2020 Frenzy Farms. All Rights Reserved.
www.krishisandhi.com

FRENZY FARMS
MARKET SURVEY
Village Sendul in UK – huge potential for multi crop production
Berni Village in Tehri, UK. Plan for Smart Farming
Understanding actual problems of farmers
Soil & Water test conducted for Smart Farming
Krishisandhi Team along with founder Subodh Shah completed the face to face Market Survey on Krishisandhi mobile app and Smart Farming concept
All these Survey conducted on farming locations only

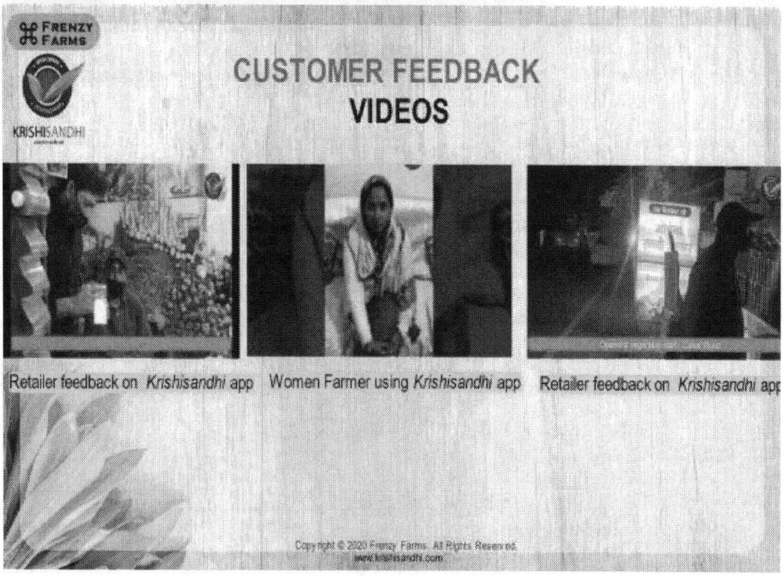
FRENZY FARMS
KRISHISANDHI
CUSTOMER FEEDBACK VIDEOS
Retailer feedback on Krishisandhi app
Women Farmer using Krishisandhi app
Retailer feedback on Krishisandhi app
Copyright © 2020 Frenzy Farms. All Rights Reserved.
www.krishisandhi.com

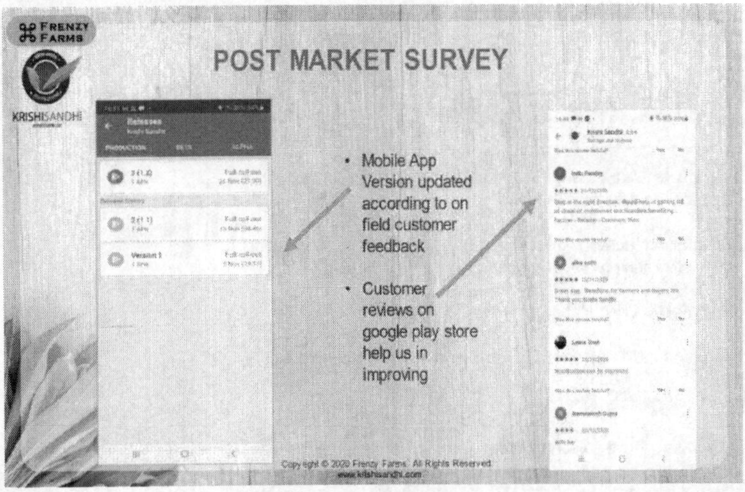

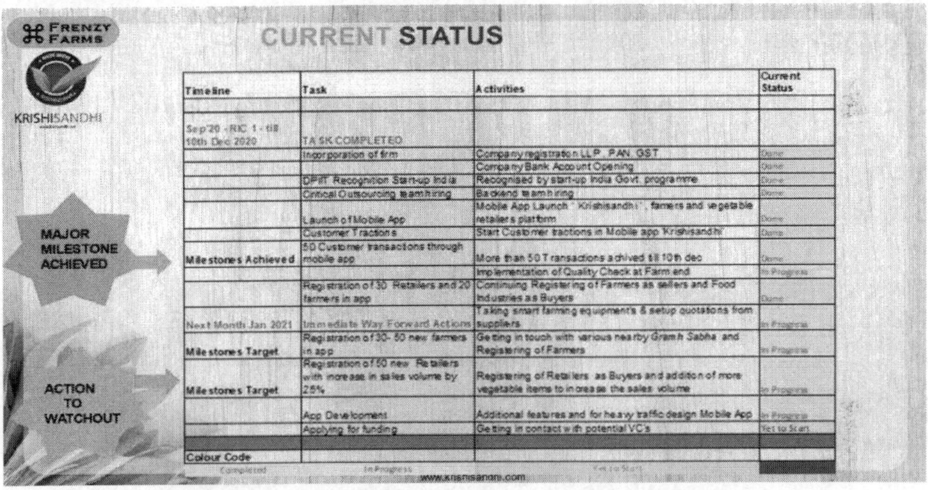

Timeline	Task	Activities	Current Status
Sep'20 - RIC 1 - till 10th Dec 2020	TASK COMPLETED		
	Incorporation of firm	Company registration LLP, PAN, GST	Done
		Company Bank Account Opening	Done
	DPIIT Recognition Start-up India	Recognised by start-up India Govt. programme	Done
	Critical Outsourcing team hiring	Backend team hiring	Done
	Launch of Mobile App	Mobile App Launch "Krishisandhi", famers and vegetable retailers platform	Done
	Customer Tractions	Start Customer tractions in Mobile app 'Krishisandhi'	Done
Milestones Achieved	50 Customer transactions through mobile app	More than 50 Transactions achived till 10th dec	Done
		Implementation of Quality Check at Farm end	In Progress
	Registration of 30 Retailers and 20 farmers in app	Continuing Registering of Farmers as sellers and Food Industries as Buyers	Done
Next Month Jan 2021	Immediate Way Forward Actions	Taking smart farming equipment's & setup quotations from suppliers	In Progress
Milestones Target	Registration of 30- 50 new farmers in app	Getting in touch with various nearby *Gramh Sabha* and Registering of Farmers	In Progress
Milestones Target	Registration of 50 new Retailers with increase in sales volume by 25%	Registering of Retailers as Buyers and addition of more vegetable items to increase the sales volume	In Progress
	App Development	Additional features and for heavy traffic design Mobile App	In Progress
	Applying for funding	Getting in contact with potential VC's	Yet to Start
Colour Code			
Completed	In Progress	Yet to Start	

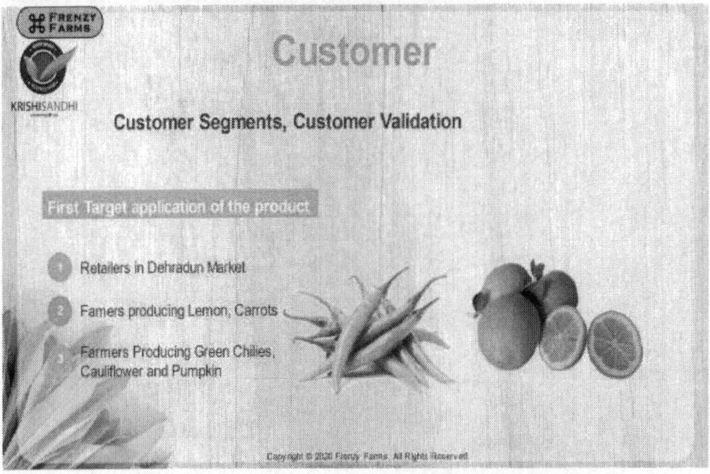

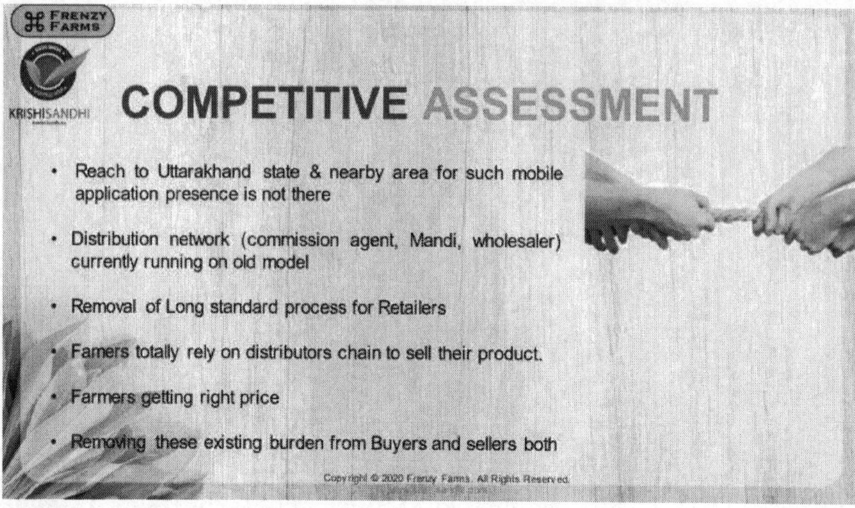
FRENZY FARMS
KRISHISANDHI
COMPETITIVE ASSESSMENT
• Reach to Uttarakhand state & nearby area for such mobile application presence is not there
• Distribution network (commission agent, Mandi, wholesaler) currently running on old model
• Removal of Long standard process for Retailers
• Famers totally rely on distributors chain to sell their product.
• Farmers getting right price
• Removing these existing burden from Buyers and sellers both
Copyright © 2020 Frenzy Farms. All Rights Reserved.

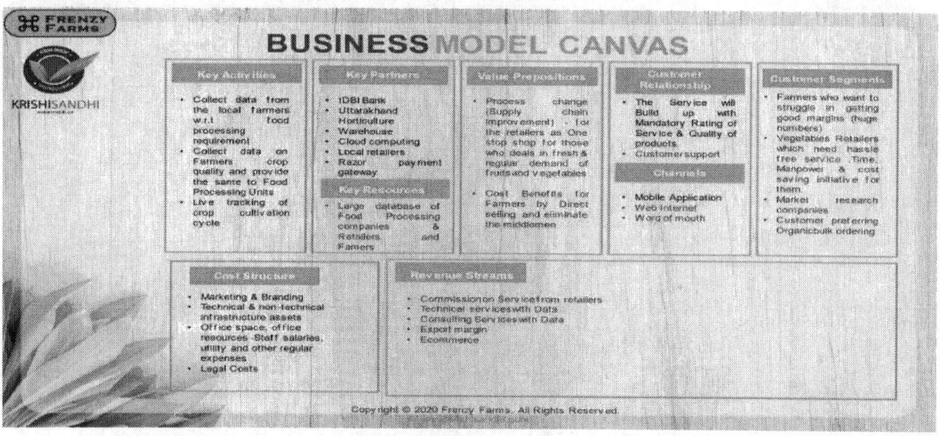
FRENZY FARMS
KRISHISANDHI
BUSINESS MODEL CANVAS
Key Activities
• Collect data from the local farmers w.r.t food processing requirement
• Collect data on Farmers crop quality and provide the same to Food Processing Units
• Live tracking of crop cultivation cycle
Key Partners
• IDBI Bank
• Uttarakhand Horticulture
• Warehouse
• Cloud computing
• Local retailers
• Razor payment gateway
Key Resources
• Large database of Food Processing companies & Retailers and Famers
Value Propositions
• Process change (Supply chain improvement) - for the retailers as One stop shop for those who deals in fresh & regular demand of fruits and vegetables
• Cost Benefits for Farmers by Direct selling and eliminate the middlemen
Customer Relationship
• The Service will Build up with Mandatory Rating of Service & Quality of products.
• Customer support
Channels
• Mobile Application
• Web Internet
• Word of mouth
Customer Segments
• Farmers who want to struggle in getting good margins (huge numbers)
• Vegetables Retailers which need hassle free service ,Time, Manpower & cost saving initiative for them.
• Market research companies
• Customer preferring Organicbulk ordering
Cost Structure
• Marketing & Branding
• Technical & non-technical infrastructure assets
• Office space, office resources -Staff salaries, utility and other regular expenses
• Legal Costs
Revenue Streams
• Commission on Service from retailers
• Technical services with Data
• Consulting Services with Data
• Export margin
• Ecommerce
Copyright © 2020 Frenzy Farms. All Rights Reserved.

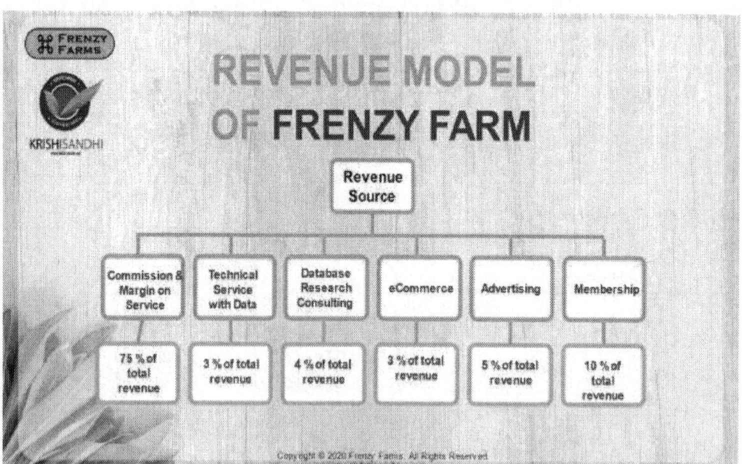
FRENZY FARMS
KRISHISANDHI
REVENUE MODEL OF FRENZY FARM
Revenue Source
Commission & Margin on Service
Technical Service with Data
Database Research Consulting
eCommerce
Advertising
Membership
75 % of total revenue
3 % of total revenue
4 % of total revenue
3 % of total revenue
5 % of total revenue
10 % of total revenue
Copyright © 2020 Frenzy Farms. All Rights Reserved.
www.krishisandhi.com

MILESTONE AND TIMELINE

Duration	Milestones	Action Points	Required Funds (inclusive of grant & owners contribution) in INR Lakh
1st Quarter	• 50 new farmers association • New Retailers to 75 • Mobile app transfer to dedicated server	Business model Qualification & Validation, , Scaling up the Retailers and Farmers chain , Farmers Training on use and benefit of mobile application*Krishisandhi*	15
2nd Quarter	• farmers association to reach 150 • Retailers to increase up to 150 • Implementation of Smart Farming model	Increase of Database with Farmers and Retailers, Smart Farming Implementatio Warehouse creations, Farmers Training on use and benefit of mobile application *Krishisandhi*	20
3rd Quarter	Retailers to increase up to 20Q50 in association with 300 farmers reach	Office setup at different locations, Farmers Training on use and benefit of mobile application*Krishisandhi*	10.3

Fund Utilization Projection

Total cost of the project: 45.3 (all figures are in INR Lakh)

Type	Expenses	Description	Q1 (40% of total grant)	Q2 (40% of total grant)	Q3 (20% of total grant)	Total Grant-in-aid	% of total grant in aid	Owner's Contribution
Non-recurring	Farm Land		Nil	Nil	Nil	Nil	Nil	10
	Machinery & Equipment	Power Weeder, Digger, Drip Irrigation	0.4	0.4	0.2	1	4 %	2
	IoT devices	Automated devices on extracting data from drip irrigations, Remote sensing services	0.4	0.4	0.2	1	4 %	2
	Mobile App Development	Dedicated server, Network services fees , & App upgradation development cost, IT support	4	4	2	10	40 %	2
Recurring	Working capital Operations	Rent Charges Office and Warehousing, Legal & Statuary Compliances	1.6	1.6	0.8	4	16 %	0.8
	Salary / Labor cost	Staff & Manpower Salary	3.0	3.0	1.5	7.5	30 %	2
	Marketing	Offers in app for Farmers & Retailers, training to farmers	0.2	0.2	0.1	0.5	2 %	0.5
	Contingency	Miscellaneous expenses	0.4	0.4	0.2	1.0	4 %	1.0
	Total		10	10	5	25	100 %	20.3

GROWTH PROJECTION

KRISHISANDHI

*all figures are approx. in ₹	Start –Up Initial Year	Year 1	Year 2	Year 3
No of Farmers	10	300-500	500-1000	1000-2500
No. of Retailers (up to)	25	250	5000	7000
No. of Farmland for Smart Farming	1	5	50	500
Expense on each Farmland	1000000	130000	100000	200000
Total Project Expenses	2000000	4500000	10000000	180000000
Total Supply in Tons X (1000kg)	10	1000	20000	30000
Commission Earning 7 per kg approx.	70000	7000000	14000000	210000000
Other revenue Earnings per farm land through Crop Data for Research (SMART FARMING)	0	0	0	10000 (Negligible)
Projected Revenue	70000	7000000	14000000	210000000
(Profit / Loss EBIT)	- 1930000	2500000	4000000	30000000

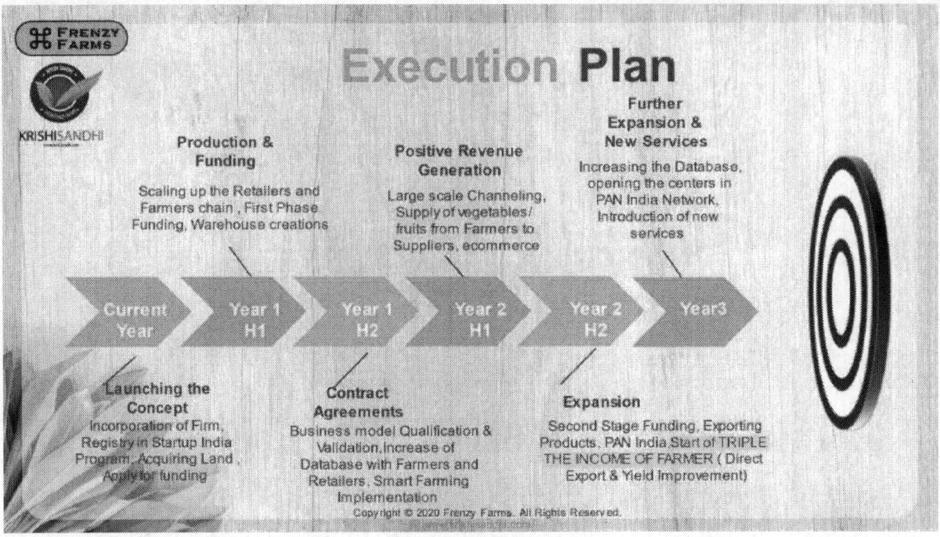
FRENZY FARMS
KRISHISANDHI
Execution Plan
Production & Funding
Scaling up the Retailers and Farmers chain , First Phase Funding, Warehouse creations
Positive Revenue Generation
Large scale Channeling, Supply of vegetables/ fruits from Farmers to Suppliers, ecommerce
Further Expansion & New Services
Increasing the Database, opening the centers in PAN India Network, Introduction of new services
Current Year
Year 1 H1
Year 1 H2
Year 2 H1
Year 2 H2
Year3
Launching the Concept
Incorporation of Firm, Registry in Startup India Program, Acquiring Land , Apply for funding
Contract Agreements
Business model Qualification & Validation,Increase of Database with Farmers and Retailers, Smart Farming Implementation
Expansion
Second Stage Funding, Exporting Products, PAN India,Start of TRIPLE THE INCOME OF FARMER (Direct Export & Yield Improvement)
Copyright © 2020 Frenzy Farms. All Rights Reserved.

FRENZY FARMS
CONTACT US
KRISHISANDHI
+91 0000000000
enquiry@xxxxxxxxx.com
Dehradun
Copyright © 2020 Frenzy Farms. All Rights Reserved.

Chapter-11

Funding Schemes (Fund your idea/Startup)

Introduction

India is an agrarian economy with agriculture contributing 17.4 per cent to the national GVA (at current prices) and providing bread and butter to 57 per cent people. It is one such domain that has always been a focus area for the government of India but remained on a low growth trajectory. The reasons may be frequent extreme weather events led by climate change, less adoption of modern agricultural technologies, lack of awareness amongst farmers for improved farming practices, absence of strong agricultural extension machinery and less number of entrepreneurial ventures and many more.

It is well established that technological progress is a key factor for agricultural growth as well as economic development. One way of achieving technological progress is by simply acquiring the technologies existing elsewhere in the world. Alternatively, nations can also attempt to enhance their own technological innovation capabilities. In case of embodied technologies, where the technology is embedded in a physical input, such as seed or machine or fertilizer, provision of safeguarding the technology through patenting is high. This also provides an opportunity for budding entrepreneurs to commercialize the ideas by marketing them.

In agriculture, generally, whenever there is a talk about entrepreneurship, an image of a rural clad appears in our mind. Now the scenario is changing. The startups, that are coming in, belong to high tech era of agriculture i.e., applications of robotics in agriculture, wearable computing and augmented reality, precision farming etc. They are young, educated and experienced beings who don't inherently belong to agriculture rather they are strong supporters of the convergence theory of different disciplines.

Even though agriculture is a complex domain dependent upon various environmental and ecological parameters, there is money to be made. Also, Government's interest in promoting the sector makes it a large pie of the cake for investors. Areas like irrigation, water, soil, warehousing and cold storages have also attracted public sector investments. Hence, Agriculture has become a mandatory topic of discussion in all startup events and investors' meets.

The focus areas in which the agricultural entrepreneurship is getting shape are Agri Inputs, Post-harvest and Food Processing, Farm Retailing, Natural Resource Management, Farm Mechanization, Precision Farming, Agri & Agri health services, ICT in Agriculture, Dairying, Biotechnology, Sustainable Agriculture, Agri Education, Supply Chain Management, Livestock and Fisheries etc.

Types of Start-up Funding

Working Capital	*Equity Financing*	*Debt Financing*	*Grants*
Brief	Equity financing involves selling a portion of a company's equity in return for capital.	Debt financing involves the borrowing of money and paying it back with interest.	A grant is an award, usually financial, given by an entity to a company to facilitate a goal or incentivize performance.
Nature	There is no component of repayment of the invested funds.	Invested Funds to be repaid within a stipulated time frame with interest	There is no component of repayment of the invested funds
Risk	Financer: There is no guarantee against his investment. Startup: Startups need to give up a portion of their ownership to shareholders.	Financer: The lender has no control over the business's operations. Startup: You may need to provide a business asset as collateral.	Financer: There is a risk of the startup not meeting the goal or objective for which the grant has been provided. Startup: There is a risk of the startup not receiving a portion of the grant due to several reasons.
Threshold of Commitment	While startups are under lesser pressure to adhere to a repayment timeline, investors are constantly trying to achieve growth targets	Startups need to constantly adhere to repayment timeline which results in more efforts to generate cash flows to meet interest repayments	Grants are distributed in different tranches w.r.t the fulfilment of the corresponding milestone. Thus, a status is constantly working to achieve the milestones laid down.

Working Capital	*Equity Financing*	*Debt Financing*	*Grants*
Return to Investor	Capital growth for investors	Interest payments	No Return
Involvement in Decisions	Equity Investors usually prefer to involve themselves in the decision-making process	Debt Fund has very less involvement in decision-making	No direct involvement in decision making
Sources	Angel Investors Self-financing Family and Friends Venture Capitalists Crowd Funding Incubators/ Accelerators	Banks Non-Banking Financial Institutions Government Loan Schemes	Central Government State Governments Corporate Challenges Grant Programs of Private Entities

Steps to Startup Fund Raising

- The entrepreneur must be willing to put in the effort and have the patience that a successful fund-raising round requires. The fund-raising process can be broken down into the following steps:
- The startup needs to assess why the funding is required, and the right amount to be raised.
- The startup should develop a milestone-based plan with clear timelines regarding what the startup wishes to do in the next 2, 4, and 10 years.
- The cost of Production, Prototype Development, Research, Manufacturing, etc should be planned well.

There are various funding agencies involved in this; Government, debt mechanism and private funding ventures. The details of these institutions are listed below:

1. **Government Grant-in-aid:** Government supports agricultural start-ups with a variety of modes. One of them is the grant-in-aid support, which is being given to the start-ups based on their eligibility and selection. Technology Development Board's grant, BIRAC's BIG Grant, NIDHI-PRAYAS, MSME etc are few names in the list of funding agencies. The entrepreneurs either have to directly apply to these agencies or they have to apply through business incubators. The grant-in-aid is generally given for product and market development. But it is a tedious and cumbersome process that takes months to execute in many cases and the ticket size is also small.
2. **Debt funding:** Generally, it is thought that only banks provide debt funding, but it is not true. Some government agencies and private investors are also providing debt assistance. Technology Development Board, Innoven Capital, Trifecta Capital etc are amongst the list of private and public-sector entities. The good news with debt capital is that the entrepreneur doesn't lose any control over the firm for raising the finances, although, raising the debt capital is a tough task.

3. **Equity Funding:** When someone establishes a company, s/he creates equity stakes (shares) adjacent to the money invested in the company. Here the start-ups are being provided with funds by charging them a percentage of their equity stakes (percentage of shares). This is how the investor becomes the shareholder of the company and is eligible to have board seat, voting rights and dividends. Although the one who invests in equity doesn't have the ambition to stay with the firm for long. They invest in the company, stay there on the board for some time till when they get their desired RoIand then they move out of the company by selling the shares.

The following entities fund the startups under Equity Funding:

a. **Government:** A 'fund of funds' of INR 10,000 crores has been established by GoI for supporting tech driven startups and agriculture is one focus area for this. The fund is being management by Small Industries Development Bank of India (SIDBI). SIDBI is investing this fund in SEBI registered Alternate Investment Funds (AIFs) and they have given 605.7 crores to 17 AIFs (Jan, 2018). Apart from this, Technology Development Board also provides equity capital to agricultural startups.

b. **Incubators:** Business Incubators also provide funding to startups from their pool of funds and charge equity against it i.e., Villgro, CIIE Ahmedabad etc.

c. **Angel Investors:** These are individual investors who have an urge to invest in startups with their surplus money. Now they are also associated with some consortium of angel investors and collectively select as well as invest in startups. As they enter at the seed stage of the company, they bear the highest risk after the founder and hence they charge handsome amount of equity from the founder upto 30 per cent. They also bring expert mentoring to the startup along with funds.

d. **Venture Capitalists (VC):** Venture Capital funds are pool of funds collected from different entities and bodies i.e., PPF, Corporate Pension Funds, Insurance Companies, HNIs, Foundations, Fund of funds etc. These are being managed by General Partners. There are many VC firms who are providing funds in agriculture i.e., Omnivore, Aspada, Ankur Capital, Unitus Seed Fundsetc.

e. **Private Equity:** Private Equity may be termed as a more mature investment where alike any other options given above, the investors invest in mature firms and buy major stakes and control of the company. They provide working capital for its diversification, expansion, new product development, new market development etc. India Value Fund is one such firm providing private equity in Agriculture.

f. **Venture debt funding:** It refers to debt funding being given to venture capital backed companies in their later stages of development. It is given to companies who have already captured a good market share, raised equity finance and are in expansion stage. Here, the companies have a decent repayment capacity and risk is also lower as compared to other stages. Many venture capital and private equity firms provide funding at this stage.

Agriculture has been a rewarding activity since ages and agribusiness is also following the same notion but it comes with its own set of challenges. Crop production and fortunes of farmer depends on weather gods to large extent. Besides this, the prototype development and validation of the offering also takes a hefty amount of time. The validation of the offering/product is necessary to comply with regulations. Unlike other areas i.e., IoT, FMCG, validation of products is on the basis of data collected from fields majorly. For this, trials need to be run on specific crops. It takes a lot of time and perseverance. Funding is a big challenge because of lack of collateral, unproven business idea and lower Return on Investment as compared to other tech sectors. Apart from this, the exit period for an investor is quite high as compared to other hot investment destinations.

Despite of all of these challenges, the investors are looking for disruptive technologies that blend traditional agricultural practices with smart solutions. They are interested in profitable ventures that witness convergence of disciplines i.e., Machine Learning (ML), Artificial Intelligence (AI), Internet-of-things (IoT) etc. They are keen on sectors that enable people to achieve highest efficiency. They believe that with better accessibility to technology, agriculture can be redefined in a better way which is being stared as the most problematic occupation signaling farmers' suicides, inefficient practices, food wastages etc. It creates a hope of revisiting the wealthy tradition of converging entrepreneurship with agriculture in a prosperous way.

Here is a list of government schemes launched to develop and encourage entrepreneurship in India.

SAMRIDH Scheme

Ashwini Vaishnaw, who was then a newly appointed Minister of Electronics Information and Technology (MeitY) launched the SAMRIDH scheme, which stands for Startup Accelerators of MeitY for pRoduct Innovation, Development, and growtH, on August 25, 2021, after a little over a week of his announcement that the government will support the startups in the initial stages.

The SAMRIDH initiative is designed to provide funding support to startups along with helping them bring skill sets together which will help them grow successful. The newly launched SAMRIDH program aims to focus on the acceleration of around 300 start-ups by extending them with customer connect, investor connect, and other opportunities for international expansion in the upcoming three years that will follow.

Startup India Seed Fund

On 16 January 2021, Prime Minister Narendra Modi announced the launch of the 'Startup India Seed Fund' — worth INR 1,000 crores — to help startups and support ideas from aspiring entrepreneurs. PM Modi said that the government is taking important measures to ensure that startups in India do not face any capital shortage.

The reserved fund for the Start-up India Seed Fund initiative, as per the Union Budget of 2022 is Rs 283.5 crore, which is higher than the revised estimate of around Rs 100 crore for the year 2021-22.

Start-up India Initiative

The Prime Minister of India launched the Start-up India Initiative in the year 2016 on 16th January. The idea is to increase wealth and employability by giving wings to entrepreneurial spirits. The government gives tax benefits to start-ups under this scheme and around 50,000 start-ups have been recognized via this scheme in a period of a little more than five years, as of June 3, 2021. The Department of Industrial Policy and Promotion is maintaining this initiative and is treating it as a long-term project. Moreover, the overall age limit for start-ups has been increased from two years to seven years. Plus, for the biotechnology firms, the age limit is ten years from the date of incorporation. It is one of the best government-sponsored start up schemes for entrepreneurs as it is providing several concessions.

ASPIRE scheme innovation entrepreneurship INDIA

The government has made continuous efforts to improve the social and economic aspects of life in rural areas of India and one of the most popular schemes that the Indian government has sanctioned in this regard is ASPIRE. A Scheme for Promotion of Innovation, Rural Industries and Entrepreneurship (ASPIRE) is a Government of India initiative and promoted by the Ministry of Micro, Small and Medium Enterprises (MSME).

The mentioned scheme was launched in 2015 to offer proper knowledge to the entrepreneurs to start with their business and emerge as employers. Since 56% of the Indian population lives in rural areas, the government has promoted entrepreneurship and innovation in the rural sector with this scheme. The ASPIRE scheme aims at increasing employment, reducing poverty, and encouraging innovation in rural India. However, the main idea is to promote the agro-business industry. The Ministry of Medium and Small Enterprises has tried to boost economic development at the grassroots level. The total budget of the scheme initially was INR 62.5 crores for the period of 2014-2016.

Pradhan Mantri Mudra Yojana (PMMY)

Mudra Bank | Government Funding Schemes for Startups in India

Micro Units Development Refinance Agency (MUDRA) banks has been created to enhance credit facility and boost the growth of small business in rural areas. The government has introduced this scheme to support small businesses in India. In 2015, the government allocated INR 10,000 crores to promote startup culture in the country. The MUDRA banks provide startup loans of up to INR 10 lakhs to small enterprises, business, which are non-corporate, and non-farm small/ micro-enterprises. MUDRA comes under Pradhan Mantri Mudra Yojana (PMMY) which was launched on 8 April 2015. The loans have been categorized as Tarun, Kishore, and Shishu. The assets are created through the bank's finance and there is no collateral security.

Ministry of Skill Development and Entrepreneurship

The task of promoting entrepreneurship was earlier given to different departments and government agencies. In 2014, the Prime Minister decided to dedicate an entire ministry to build this sector as he felt that skill development required greater push from the government's side. Furthermore, the idea is to reach 500 million people by the year 2022 through gap-funding and skill development initiatives.

ATAL Innovation Mission

Atal Innovation Mission | Initiatives for Entrepreneurship development

In the budget session of 2015, the Indian government announced the Atal Innovation Mission (AIM); with the name coming from Atal Bihari Vajpayee, the Former Prime Minister of India. Atal Innovation Mission was established to create a promotional platform involving academicians and draw upon national and international experiences to foster a culture of innovation, research, and development. The government allocated AIM around INR 150 crores in the year 2015.

Dairy Processing and Infrastructure Development Fund (DIDF)

National Bank for Agriculture and Rural Development (NABARD) is an apex development bank in India. The Government of India announced the creation of the Dairy Processing and Infrastructure Development Fund under NABARD in the Union Budget of 2017-18 for the sustained benefit of farmers. The total corpus for this fund is INR 8000 crores over a period of 3 years (i.e. 2017-18 to 2019-20)

Milk Unions, multi-state milk cooperatives, state dairy federations, milk-producing companies, and NDDB subsidiaries meeting the eligibility criteria under the project can borrow loans from NABARD. The loan component would be 80% (maximum rate) with the end borrower's contribution at 20 % (minimum rate). Borrowers shall get the loan at an interest rate of 6.5% per annum. The period of repayment will be 10 years. The respective state government will be the guarantor of loan repayment. Moreover, if the borrower is not able to contribute his or her share in the scheme, the state government shall step in.

The departments that come under the agriculture ministry were allocated a total of Rs 1.31 lakh crore in 2021 Budget, which has been increased to Rs 1.32 lakh crore in the Union Budget of 2022. A fund consisting of blended capital that will be raised under the co-investment model will be facilitated via NABARD, which will finance the agritech startups and rural enterprises that are relevant to farming.

The Finance Minister of India, Nirmala Sitharaman further added that the use of 'kisan drones will see a new encouragement to facilitate effective crop assessment, digitization of land records, and the spraying of insecticides and nutrients.

Support for International Patent Protection in Electronics & Information Technology (SIP-EIT)

The Department of Electronics and Information Technology (DeiTY) has launched a scheme entitled "Support for International Patent Protection in E&IT (SIP-EIT)". This scheme provides financial support to MSMEs and Technology Startups for international patent filing.

Features and benefits of the SIP-EIT scheme are:

Financial support is provided for international filing in Information Communication Technologies and Electronics sector.

The Reimbursement limit has been set at a maximum of INR 15 lakhs per invention or 50% of the total charges incurred in filing and processing of a patent application, whichever is lesser.

The SEP-EIT scheme can be applied at any stage of international patent filing by the applicant.

MGS | Government Grants for Start-ups in India

Department of Electronics and Information Technology (Deity) started the Multiplier Grants Scheme (MGS). This scheme aims to encourage collaborative Research & Development (R&D) between industry and academics/institutions for the development of products and packages. Under the scheme, if the industry supports the R&D of products that can be commercialized at the institutional level, the government shall provide financial support which will be up to twice the amount provided by industry. MGS promotes and expedites the development of aboriginal products and packages. The government grants would be limited to a maximum amount of INR 2 crores per project and the duration of each project could considerably be less than 2 years. It would be INR 4 crores and 3 years for industry associations.

Credit Guarantee Fund Trust for Micro and Small Enterprises (CGTMSE)

The Credit Guarantee Fund Trust for Micro and Small Enterprises (CGTMSE) was set up by the government of India and had been put to effect from 1st January 2000 onwards to provide business loans to micro-level businesses, small-scale industries, and start-ups with zero collateral. It allows businesses to avail loans at highly subsidized interest rates without requiring security. By working along with SIDBI (Small Industries Development Bank of India), the government provides a maximum amount of up to INR 100 lakhs under this scheme for boosting new enterprises as well as rehabilitating the existing ones. Primarily meant for manufacturing units, this loan can be availed in the form of working capital or a term loan.

Software Technology Park (STP)

The Software Technology Park (STP) scheme is a totally export-oriented scheme for the development and export of computer software. This includes the export of professional services using communication links or media.

The scheme is unique in its nature as it focuses on only one sector, i.e., computer software. The scheme integrates the government concept of "100% Export Oriented Units" (EOU), "Export Processing Zones" (EPZ), and the concept of Science Parks or Technology Parks as operating elsewhere in the world. The sales in the Domestic Tariff Area (DTA) shall be permissible up to 50% of the export in value terms. STP gives total depreciation on capital goods over a period of five years.

The Venture Capital Assistance Scheme (VCA)

Small Farmer's Agri-Business Consortium (SFAC) has launched the Venture Capital Assistance (VCA) scheme for the welfare of farmer-entrepreneurs and to develop their agri-business. The scheme is approved by the banks and financial institutions regulated by the RBI. It intends to provide assistance in the form of term loans to farmers so that the latter can meet the capital requirements for their project's implementation. VCA promotes the training and nurturing of agri-entrepreneurs.

The quantum of the loan will be 26% (40% for hilly regions) of the promoter's equity. The maximum amount of loan provided under this scheme will be INR 50 lakhs.

NewGen Innovation and Entrepreneurship Development Centre (NewGen IEDC)

NewGen IEDC is an initiative launched by the National Science and Technology Entrepreneurship Development Board under the Department of Science and Technology, Government of India. The initiative aims to inculcate the spirit of innovation and entrepreneurship among the Indian youth. It also endeavors to support and encourage entrepreneurship through guidance, mentorship, and support.

NewGen IEDC is a five-year programme that would be implemented in educational institutions. It will support up to 20 new projects.

Single Point Registration Scheme (SPRS)

The Single Point Registration Scheme (SPRS) was launched in 2003. It is managed by the National Small Industries Corporation (NSIC). NSIC registers all Micro & Small Enterprises (MSEs) in India under the Single Point Registration Scheme to enable them to participate in government purchases.

Enterprises are classified as Micro, Small, or Medium based on the limit of investment. Eligible MSME units are provided with Udyog Aadhar registration certificate. All central ministries, departments, and PSUs shall set an annual goal of minimum 20% of the total annual purchases of products produced or rendered by MSMEs. About 358 items are reserved for exclusive purchase from MSMEs.

Modified Special Incentive Package Scheme (M-SIPS)

The government of India has approved a special incentive package to promote large-scale manufacturing in the Electronic System Design and Manufacturing (ESDM) sector. The scheme is called the Modified Special Incentive Package Scheme (M-SIPS).

Under M-SIPS, the Indian government will provide a subsidy of 20% on capital investments in special economic zones (SEZs) and 25% on capital investments in non-SEZs for individual companies. It also provides the re-imbursement of CVD/ excise on capital equipment for non-SEZ units. Re-imbursement of central taxes and duties is also provided for high technology and high capital investment units.

Stand Up India Scheme

Stand-Up India for Financing SC/ST and/or Women Entrepreneurs

Stand-Up India is a notable government of India scheme for financing SC/ST and/or women entrepreneurs. According to the scheme, bank loans between 10 lakhs to 1 crore can be borrowed by at least one Scheduled Cast (SC) or Scheduled Tribe (ST) borrower and at least one woman per bank brand to set up a greenfield enterprise. The greenfield enterprise may be based out of manufacturing, services or the trading sector. In case of non-individual enterprises, it is mandatory that an SC/ST or a woman entrepreneur holds at least 51% of the shareholding and controlling stakes.

ExtraMural Research or Core Research Grant

The government of India launched the famous Extramural Research (EMR) funding scheme via SERB to help academic institutions, research laboratories, and other R&D organizations carry out their basic research in all frontier areas of Science and Engineering. Founded just after the establishment of the Science and Engineering Research Council (SERC), this scheme is one of the oldest on the list and still quite a relevant one even after more than four decades of its existence.

This scheme encourages emerging and eminent scientists in the field of science and engineering for an individual-centric competitive mode of research funding. Only the name Extramural Research (EMR) has now been renamed to Core Research Grant (CRG) because the scheme provides core research support to active researchers.

High Risk - High Reward Research

The High Risk and High Reward Research is a scheme launched by the Indian government to support and invite new proposals and ideas that have the potential to usher a paradigm shifting influence on the Science and Technology domains. This funding focuses on the new proposals, which might be conceptually new and risky but are expected to have a paradigm shifting influence on the S&T, in terms of formulating new hypotheses or scientific breakthroughs, which might help in the emergence of new technologies.

IREDA-NCEF Refinance Scheme

The revival of the operations of existing biomass power and other small hydro power projects, which were affected due to unforeseen circumstances, is essential, which is why the government of India decided upon the Refinance Scheme in collaboration with the Indian Renewable Energy Development Agency Ltd. (IREDA) and The National Clean Energy Fund (NCEF).

The scheme aims to revive the operations of the existing biomass power and small hydro power projects by cutting down the cost of funds for these projects and providing refinance at concessional rates of interest, with funds sourced from the National Clean Energy Fund (NCEF).

Dairy Entrepreneurship Development Scheme

The Dairy Entrepreneurship Development Scheme (DEDS) is a scheme implemented by the Department of Animal Husbandry, dairying, and fisheries to generate self-employment opportunities in the dairy sector. These opportunities in the mentioned sector will comprise activities like the enhancement of milk production, procurement, preservation, transportation, processing, and marketing of milk by providing back-ended capital subsidy for bankable projects.

The scheme has been implemented by the National Bank for Agriculture and Rural Development (NABARD).

Zero Defect Zero Effect (ZED) Certification Scheme

The Zero Defect Zero Effect (ZED) scheme has been launched by the government of India with a vision of creating proper awareness about ZED manufacturing among the MSMEs and motivating them for the assessment of their enterprise for ZED and supporting them. ZED can be summed up as an integrated and holistic certification and handholding scheme that extends an opportunity to the Micro, Small and Medium Enterprises (MSMEs) to strive to continuously improve their processes and move up the ZED maturity assessment model.

Credit Linked Capital Subsidy for Technology Upgradation (CLCSS)

Credit Linked Capital Subsidy for Technology Upgradation, which is popularly acronymed as CLCSS, is a government of India scheme to upgrade technology. Under CLCSS the Indian government provides cash for the Indian companies upfront to upgrade/modernize their equipment or techniques. This scheme mainly empowers Micro and Small Scale Enterprises (MSMEs).

Design Clinic Scheme for Design Expertise

The Design Clinic Scheme for Design Expertise is a scheme declared by the Indian government to support the MSME manufacturing sector of India. As the government deems that design and innovation are critical to the growth of a brand and feels that the MSMEs should develop a design-centric approach to fuel their startups, it aims to infuse design expertise in them. Under this scheme, the government of India announced to extend around Rs 60,000 for attending design seminars and up to Rs 3.75 lakhs or 75% of the cost that would be needed in conducting the seminar, where the entrepreneurs and their teams can learn about design theories, interact with design veterans, build a network, and put them into practice.

Chapter-12

Myths in Entrepreneurship

Myths about Entrepreneurs That You Should Know Before You Start Your Business and Convert Myths into a source of Motivation to take up Start-up/ Business as Career Option"

Around one out of every 18 persons in the globe owns a business. Some people start businesses to seek opportunities, while others do it out of need, such as many entrepreneurs in poor countries, since it's the most promising way to make a livelihood in their area. Every would-be entrepreneur can benefit from understanding what the role requires, regardless of their location or motive.

The many myths and misunderstandings that surround entrepreneurship are a good place to start; refuting them can assist start-up leaders figure out what's genuinely required to succeed with their young business. The following list shows why some of the most frequent beliefs about entrepreneurs are incorrect.

1. Entrepreneurs are born that way: Many people believe that entrepreneurs are born with specific innate talents and that only those with certain talents can be entrepreneurs. The truth is that practically anyone who can gain the requisite skills can become an entrepreneur. True, some people acclimate to the demands of the job faster than others, but there's no rule that says only certain people can start businesses. Entrepreneurs can be gregarious extroverts or quiet introverts; they can be "big picture" thinkers or detail-oriented when it comes to executing a business plan. Entrepreneurship is an acquired skill rather than a natural talent.

2. The only requirement is a good idea. Many people also assume that all it takes to achieve the American dream is to come up with one brilliant business concept. While the idea isn't wholly incorrect, it is deceptive. Even the most brilliant ideas, especially those with the potential to revolutionise an entire industry, require adequate execution to become a reality. Ideas are crucial, but so are talent, leadership, communication, and a variety of other elements.

1. Starting a new business guarantees freedom.

Many people are drawn to the possibility of starting their own business because it allows them to break free from the standard 40-hour work week. Many people discover that, while leaving behind their old schedules and creative restrictions, they are replaced with new demands. Sure, there will be more freedom in some ways, but entrepreneurship frequently necessitates significant sacrifices. It has the potential to dominate every waking moment of your life; the job doesn't stop when the clock strikes 5 p.m.

2. Launching a company quickly leads to wealth.

Some entrepreneurs make the mistake of thinking that starting a business will put them on the fast route to making a lot of money — quickly. While some businesses succeed right away, others take a little longer to get there. One of the most difficult challenges for an entrepreneur is to properly time the company's expansion and to maintain growth.

3. Only money motivates employees.

The belief that money is the best method to encourage employees is a misconception associated with the "get rich quick" concept. According to recent statistics, Millennials, who are increasingly becoming the majority in the workplace, would rather earn less at a job they enjoy than earn more at a job they despise. To overcome this misunderstanding, the entrepreneur must build not simply a revenue-generating business plan, but also a company culture that engages people.

4. Businesses either flourish or fail.

It's easy to adopt an all-or-nothing attitude toward entrepreneurship because only a few startups that explode out of the gate receive the majority of the press. The high failure rate for new enterprises may appear to support this theory. Entrepreneurs must understand, however, that getting a business off the ground and keeping it afloat requires a lot of patience and forethought. Some businesses that flounder or have slow growth rates at first go on to enjoy substantial growth later on.

6. All responsibility falls on the entrepreneur.

Another myth about starting a business is that it is entirely dependent on the entrepreneur. This may be true in the beginning, but adopting this concept too seriously is the surest way to burn out. Collaboration and delegating are important aspects of a company's health. Nobody can accomplish it all by themselves.

7. There's a secret, "silver bullet" key to success.

Many successful businesspeople give the impression that they've discovered some type of secret formula for success. This, however, ignores the entrepreneur's previous failed business ventures, their old-fashioned hard work and patience, or any of the many other elements that go into building a strong company. The truth is that there is no single key to success. Entrepreneurial success, if anything, necessitates a combination of many ideas, people, and resources that must all come together at the right moment and in the right location.

8. Businesses need someone with an MBA at the helm.

When it comes to entrepreneurs and business degrees, conflating value with necessity is a mistake. On the one hand, a formal education has incalculable advantages. Entrepreneurs, on the other hand, are not required to hold an MBA or other business degree by the market. Some company founders, for example, have degrees in engineering and use their in-depth technical knowledge to uncover technology gaps and design solutions. Other successful start-up founders do not have a college diploma.

9. Quitting is for losers.

One final misunderstanding is that entrepreneurs must put everything on the line, no matter what. But the reality is that not every idea will turn into a viable business. Successful entrepreneurs frequently cycle through and test a variety of concepts before settling on one that sticks. Quitting may appear to be a failure, but it's actually a frequent aspect of the entrepreneur's journey and can teach you a lot. It's crucial to know when to walk away and move on to the next idea.

Annexure-I

ENTREPRENEURSHIP CASE STUDIES

Entrepreneurship and entrepreneurial education are considered two important and determinative issues in the advancement of societies. In reality, both concepts need to be interconnected, as they are the mechanisms for boosting the bet on innovation, creativity, employment and economic growth. Therefore, entrepreneurship education must be encouraged, and true stories of entrepreneurs can be examples to young people to help them in motivation and taking the risk with ease by adopting strategies.

Case study-I

Cost Effective Shade Net House

Name: Shri Jaganath

Profile: Shri Jagnnath Gangaram Tayade

A/P Ladsavangi, Taluq & District Aurangabada, Maharashtra

(Mobile: 09421313616)

Age: 50 years

Education: 10 std

Landholding: 7.2 ha

Farming experience: 27 Years

Crops grown: Cotton, maize, sweet orange seed production of capsicum, tomato and water melon

Livestock: Cow and buffaloes

Description of innovation

Shade net house was developed by Shri. Jagnnath Tayade from locally available iron material. Total height of the shade net house is 6m from center place, Length is 36 m and width is 24 m. The shade net house requires wire ropes. Here the obstacles within shade net are negligible which helps for smooth intercultural operations and easy movement of labour as compared to recommended one.

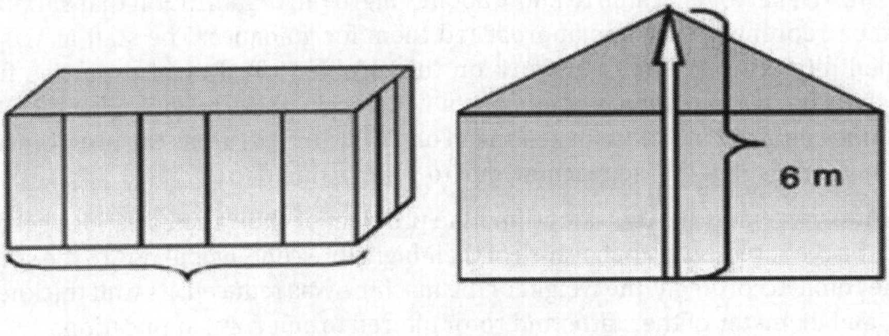

Cost effective shade net house

Practical utility of innovation

This type of shade net house is suitable for growing heighted crops like capsicum, tomato, chili, etc. This shade net house sustains wind speed up to 80 km/hr and is economically viable as the total cost required to construct is Rs 60000 which become 30% less than the recommended technology and rain water is drained easily due to doom shape structure. At present 15 farmers had adopted this type of structure on 0.10 ha each and are taking seed production programme of capsicum and tomato

Source: AME Foundation

CASE STUDY-II

An innovation that generates revenue for rural women

Mr. K. Vivekanandan of Coimbatore (Tamil Nadu) invested Rs.8 lakh and created a 3 HP pin pulverizer for grinding chilli and coriander. "The machine is an ideal revenue generator for rural women who are interested in increasing their domestic income," says Mr. Vivekanandan.

Most of the existing machines to grind chilli and coriander require high installation costs, and use a lot of power making it unsuitable for rural areas, where power supply may also be undependable.

Challenges faced

Mr. Vivekanandan thought he solved 90 per cent of the grinding problem when he developed the machine, and manufactured nearly 100 of them. But to his dismay he found that only 20 pieces found buyers. Some of the buyers returned the machine, because the chilli and coriander did not pass through the filter screen, and created too much dust while grinding.

Work came to a standstill, and did not resume for almost a year.

Mr. Vivekanandan came to know about Villgro (an organization that supports rural entrepreneurs) and he approached them for guidance. The staff at Villgro tapped different resources to work on this problem. Technical expertise first assisted Mr. Vivekanandan in producing a 1 HP, single phase machine, because the machine could not initially run at a speed on 3 HP. (In rural areas the preference is for a one- HP, single-phase machine due to voltage fluctuation).

After several trials they identified the problem of chilli and coriander getting stuck on the screen was not because of their high fibre content, but due to the speed of the rotor. Accordingly, the weight of the machine was reduced, its wall thickness, size, and diameter of the stator and rotor altered to suit rural application.

Cost

Mr. Vivekanandan brought down the cost of the machine to cater to rural needs by focusing on the types and amount of materials used. Each unit is priced at Rs. 11,500 (with motor).

Case Study-III

Name of the Startup: BB&GG(Be Blessed & Go Green), Srinagar, Jammu and Kashmir.

Elevator Pitch: Making organic fertilizers from human hair and Chicken feathers which would otherwise go to the landfills.

Founder(s): Ahmer Bashir (28), Mohsin Bashir (31)

Established: LLP, 2020.

Problem statement: Despite making up a huge proportion of the municipal waste, there is no waste management plan for keratinous waste that takes more than 50 years to decompose.

The solution: Convert the hair and feathers found in poultry waste into agricultural fertilizers.

THE STORY: Being from a business family, Ahmer wished to go into business himself. However, as he grew up, he developed fondness for science and started considering a career in research. Eventually, he has managed to fulfill both those ambitions as a co-founder of a startup that is marketing a product devised by his team for which they have also received a granted patent from Indian Patent office.

After he finished Masters in Agricultural Engineering & Technology, he came to know about the sheer amount of keratinous waste which got mixed with the garbage that reaches the landfills. In his hometown of Srinagar alone, more than 700 kgs of human hair wastes and 1.80 lakh kgs of poultry wastes are generated each month. He soon found out that around 125 million kgs of human hair waste from salons, temples and 350 million tons of chicken poultry wastes are generated in India annually. Thehard keratinous waste products take more than 50 years to decompose, making it an severe environmental concern.

Meanwhile, as a student, Ahmer had learnt how overuse of chemical fertilizers have resulted in destroying the soil quality, and subsequently, the yield of Indian agricultural holdings. It got thinking of ways to utilize the keratinous waste into products. Professor Imtiyaz Murtaza from Division of Basic Sciences & Humanities mentored and directed Ahmer in several attempts to finally devise a method to convert human hair and poultry feathers into organic fertilizer in just 50 minutes. The resultant liquid is Rich in several minerals and micronutrients, while being cost effective and completely natural to be used in Indian Agricultural Systems safely.

Success mantra: Making use of waste products available abundantly and readily provided by the civic bodies as the ingredients of organic fertilizer.

Future plans

Find more uses for the keratinous waste including utilizing it as a feed. Making the product available to farmers all over the country.

Annexure-II

Detailed Project Report for Start-up

Detailed Project report Format for New Business Start-up Ventures

Below is the sequence of standard format which should be followed while preparing new business project report:

1. Background of the business
2. Customer's profile
3. Long- and short-term Corporate Objectives
    - To perform a viability assessment of the proposed new business ideas in terms of marketability, technical feasibility, financing and authorities
    - To be able to prepare a relevant business plan
    - To recognize fundamental startup issues
4. Market Analysis
    - Brief discussion on the type of market, chief influencers, players, etc
    - Market description
    - Reasons for starting business in a particular market
    - Target clients
    - Advantages of the services offered by the new business
    - Market consumption patterns
    - Past and existing supply location
    - Production prospects and limitations
    - Exports and Imports
    - Price structure
    - Flexibility of demand
    - Client behaviour, purposes, intentions, impetus, approaches, inclinations and needs

- Supply network and marketing rules formulated by the government
- Government and technical limitations imposed on the promotion of the product

5. Financial Assessment
    - Investment expenditure and value of the entire project
    - Methods of investment
    - Anticipated productivity
    - Money flows of the project report
    - Investment value evaluated in context of different points of merit
    - Estimated financial ranking
6. Marketing Assessment
    - Product
    - Price
    - Place
    - Promotion
7. Operational Plan
    - Business models
    - Production of goods and services
8. Financial Plan
9. Management Structure
10. Business structure (Ownership, staff, etc)
11. SWOT Analysis
    - Significant Success aspects depending on Strengths, Weaknesses, Opportunities and Threats to be faced by the firm in future
12. Appendices
    - Break-Even Assessment
    - Profit and Loss Synopsis
    - Fund Flow Summary

Designing an effectual Project report for New Business

While designing a project report for new business it is vital to keep the following points in consideration:

1. Target Market:

While making a new business project report it is significant to keep in mind the target market. For instance, if you are preparing a report on debt financing, the focus should be on the assurance that the debt can be reimbursed by the company and not on the huge revenues that the firm has accrued.

2. Market approach:

The first approach towards preparing business report should be developing and assisting powerful business strategy. The report should describe the economic trends, clients and players. It should also explain on how to perform the chosen approach, the marketing of its products and functional competence.

3. Market Rivalry:

An industrialist should recognize at what point of time the firm will perform similar activities and when it will do things another way, in order to identify its strength and drawbacks. Concentrate on strategies which present you differently from your competitors, market products differently using various channels, etc.

4. Be practical:

Catalog realistic plans, for there are many unanticipated expenses, price overruns, issues that have been neglected, etc. Hence, predict future trends realistically to make a powerful project report.

Annexure-III

Successful Start-ups in Agriculture

The agriculture sector in India is beset by issues such as outdated equipment, inadequate infrastructure, and farmers who are unable to easily access a larger range of markets while making only marginal returns on crop sales. The most important challenges are infrastructure and supply chain management. With significant advancements in technology, digitization, and start-up culture, many new ones are entering the Indian start-up ecosystem.

The main aim of enlisting the Successful Agri Startups in this Toolkit is to provide new budding innovators a right information and motivational source for these start-ups to take up entrepreneurship as career option.

SFarmsIndia - Platform for Listing & Fraction Trading of Agri Lands

Agriculture Startup	*SFarmsIndia*
Founded In	2018
Headquarters	Hyderabad
Founders	Kamesh Mupparaju, Linus Lindgren
Total Funding	$50k

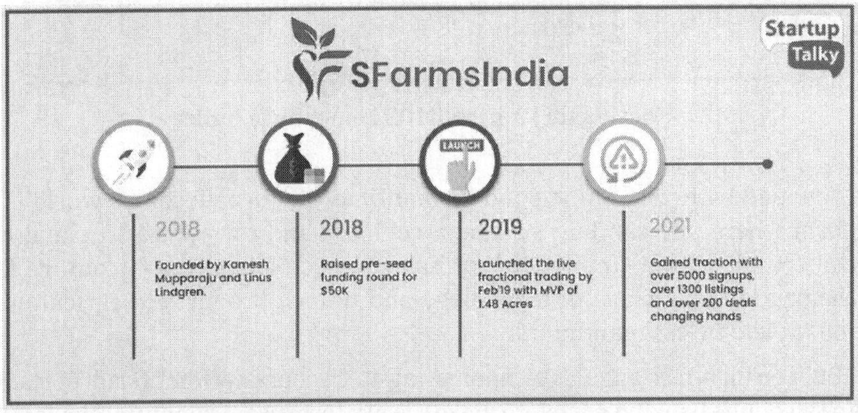

SFarmsIndia - Agriculture Startup in India

Founded by Kamesh Mupparaju and Linus Lindgren in 2018, SFarmsIndia is an online web application with two core functions; listing and fraction trading. These two functions work together creating a great synergy. On the listing platform buyers can find/search agriculture lands and sellers can list/post the agriculture lands for sale/lease. On the other hand, SFarmsIndia's main USP is that it makes an agreement with the seller (Agri-realty developer) to issue fractions by digital land contract against their lands. Each agriculture land is assigned with a digital contract and divided into 2000 fractions per acre, thereby enabling medium to small income groups to acquire lands easily. After owning the fraction, people can sell it directly through the trading platform.

Technology is the core of SFarmsIndia's business operations. Using technology, they intend to create an effective and efficient platform for Agri-lands that can eliminate the middlemen who inflate the prices.

KhetiGaadi - Marketplace for Tractors & Farm Machinery

Agriculture Startup	Khetigaadi.com
Founded In	2016
Headquarters	Pune
GMV	INR 1 Crore (2020)
Founders	Pravin Shinde and Vishnu Dhas

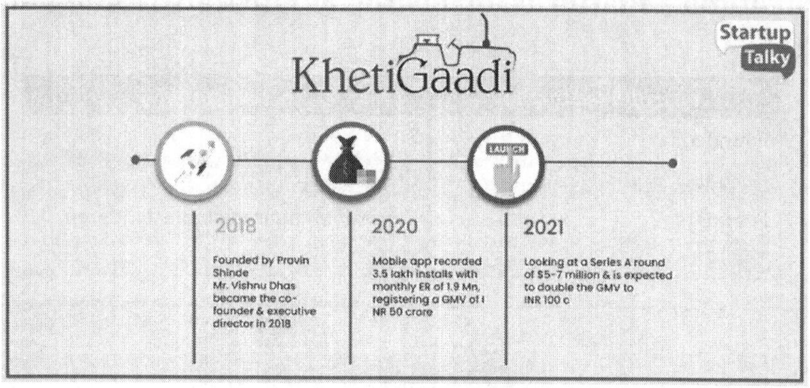

KhetiGaadi | Agritech IOS Company in India

Founded by Pravin Shinde and Vishnu Dhas, Khetigaadi.com is world's first platform where one can buy, sell, rent, compare and review tractors and farm machinery, all with a simple click. It has been established to encourage farm mechanization and usage of technology and hi-tech products in agriculture to optimize yield in less man-hours.

Built on high-tech angular technology, agritech company KhetiGaadi portal is as safe as an online banking portal. To make the platform user friendly for the farmers,

KhetiGaadi has made it available as an App in 10 languages and the company's website can be viewed in 3 different languages, English, Hindi and Marathi. The concept is techno-savvy and compatible with iOS as well as Android. As of 2021, this agritech startup is looking to raise Series A round & expected to double the GMV to INR 100 crore.

Farm2Fam - Redefining Health via Microgreens

Agriculture Startup	Farm2Fam
Founder	Keya Salot
Founding Year	2019
Headquarters	Mumbai
Parent Organization	Vijay Transtech Pvt. Ltd.

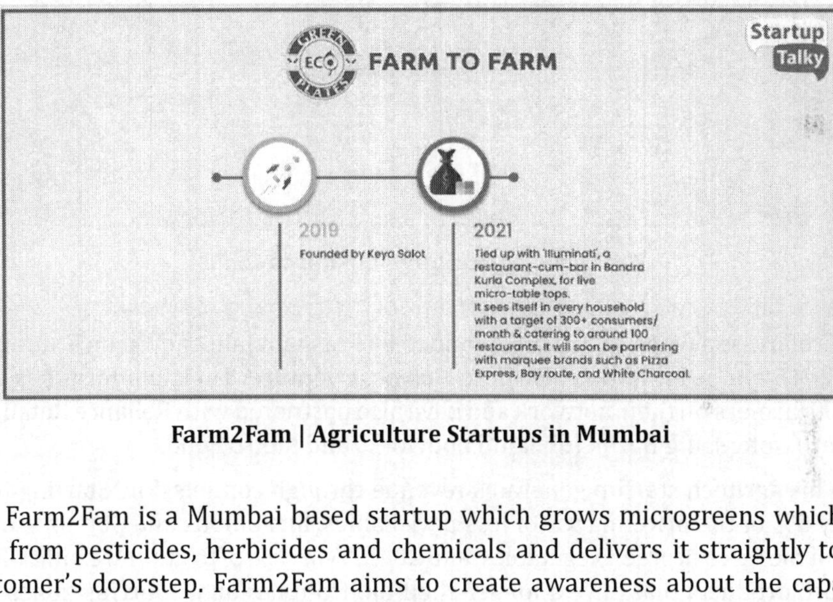

Farm2Fam | Agriculture Startups in Mumbai

Farm2Fam is a Mumbai based startup which grows microgreens which are free from pesticides, herbicides and chemicals and delivers it straightly to the customer's doorstep. Farm2Fam aims to create awareness about the capacity of the human body to heal itself with natural nutrition. Their intention to use technology in combination with traditional Indian agricultural methods to grow niche nutritional products.

Farm2Fam was founded in January 2019 by Keya Salot. She is a law graduate and worked with reputed law firms prior to starting Farm2Fam. This Microgreen-based agritech startup is soon going to partner with marquee brands like Pizza Express, Bay Route and White Charcoal. It sees itself in every household with a target of 300+ consumers per month.

Crofarm - Farm to Business Venture

Agriculture Startup	Crofarm
Founded In	2016
Headquarters	Delhi
Founders	Varun Khurana, Prashant Jain
Total Funding	$3.783 Million
Investors	Vinay Mittal, Ramit Sethi, Pravega Ventures, FactorE Ventures, Ashish Chand

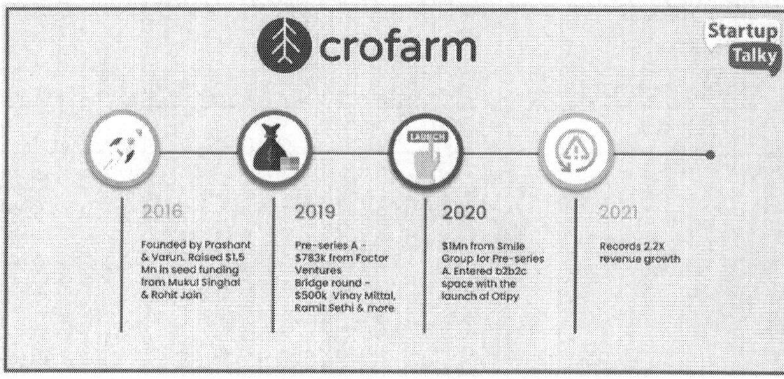

Crofarm | Agriculture Startups in India

Crofarm is a website that was founded by Prashant Jain and Varun Khurana in May 2016 and is an F2B i.e. Farm to Business venture. Their platform has over 10,000 farmers on their network and have also partnered with Reliance Retail, Big Basket, Grofers, Big Bazar, Jubilant Foodworks and Metro Foods.

This agritech startup generates revenue through commission. Starting from nearly 5% of the price in case of less perishables, like onions and potato, around 15% of the price of green vegetables, and 20-25% in case of fruits and exotics. They procure products that have a longer shelf life from national sourcing zones and other products from regional sourcing zones. This agriculture startup has raised $1.5 million in seed funding from investors like Mukul Singhal and Rohit Jain, former Principals at SAIF Partners and recently raised $1mn from Smile Group for pre-series A round and entered B2B2C space with the launch of Otipy in 2020.

BOHECO - Reinventing Industrial Hemp-based Products

Agriculture Startup	Bombay Hemp Company
Founded In	2013
Headquarters	Mumbai
Founders	Avnish Pandya, Chirag Tekchandaney, Delzaad Deolaliwala, Jahan Peston Jamas, Sanvar Oberoi, Sumit Shah, Yash P. Kotak

Revenue	INR 7 Crore (2020-21)
Investors	Ratan Tata, Rajan Anandan, Sanvar Oberoi, Srinivaas Sirigeri, Nikhil Velpanur, Yash Kela, GINI Intl Ltd.

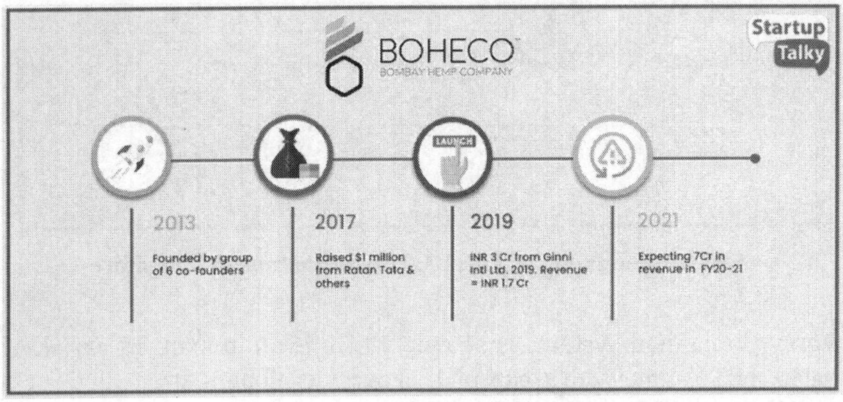

Boheco | Farming Startups in India

BOHECO – Bombay Hemp Company – is a Mumbai-based social enterprise founded in 2013 to research on and promote industrial hemp. Their operations range from research, cultivation, harvesting, processing, manufacturing, trading, wholesaling, retailing, innovating, creating a market for and promoting Indian industrial hemp both across the country and around the world.

Boheco uses public- private partnership, contractual farming and processing, subsidiaries and collaborative research models with the aim to build an industrial hemp ecosystem. Boheco is an agro-based startup that is reinventing the future of agriculture and sustainable living with hemp as their lens in India. The startup harnesses the power of agriculture by advocating industrial Hemp and its benefits to society. They give assistance to local farmers in cultivating their crops by providing optimum seeds, best practices in cultivation techniques and innovating the Agro-products.

Aarav Unmanned Systems - Building the Future of Drones

Agriculture Startup	Aarav Unmanned Systems
Founded In	2013
Headquarters	Bengaluru
Founders	Nikhil Upadhye, Suhas Banshiwala, Vipul Singh, Yeshwanth Reddy
Investors	BellWether Advisors, 500 Startups, 3one4 Capital, Sanjay Jesrani, The Phoenix Fund, among others

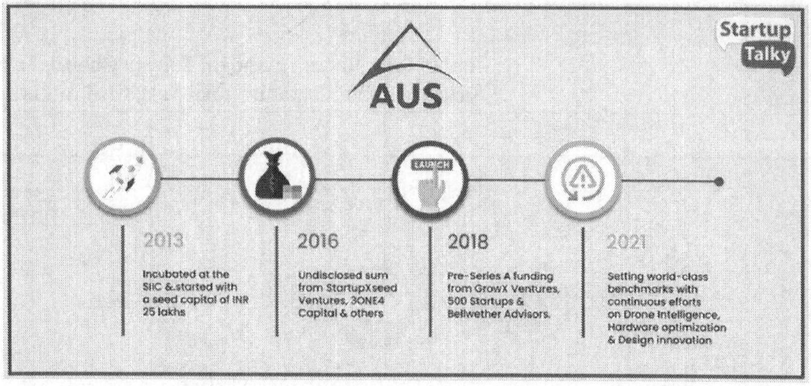

Aarav Unmanned Systems | Agritech Startups in Bangalore

Aarav Unmanned Systems is startup which is an unmanned aerial vehicle incubated at IIT Kanpur with a team of designers, developers, artists, and engineers from IIT Kanpur, IIT Bombay, IISC Bangalore & IIST. They aim to build the future of drones and their applications in the enterprise space. The company's drones provide high-value engineering solutions to enterprises across geographic information system surveying/mapping, industrial inspection and precision agriculture. Thus, it holds the potential for making precision agriculture scalable in the nation, which is far behind in employing this type of technology.

They provide accurate 3D representations of the terrain surface and it can transform and change the way decisions are made in agriculture through its solutions to optimize irrigation, fertilization, pesticide distribution and early failure warnings. This Agritech Startup's recent funding came in 2018 as a pre-series A round from GrowX Ventures, 500 Startups and Bellwether Adviors.

Aibono - Leverages Data Science to increase Farmers' Yields

Aibono | Agriculture Startups in India

Agriculture Startup	Aibono
Founded In	2014
Headquarters	Banglore
Founders	Vivek Rajkumar
Total Funding	\$5 Million
Investors	Ashish Kacholia, S Chand Group, Menterra Ventures, Redbright Partners and others

Aibono is an agritech founded by IIT Madras graduate Vivek as Airwood Aerostructures, and rebranded as Aibono, which is a company that provides farming related intelligence, technology, expertise and gadgets to farmers. The startup advices farmers on the right amount of inputs to use for maximum yield. They began in the niche area of providing Farm Management-as-a-Service whereby, a farmer gets to outsource his entire measurement, production management and decision-making processes to a Service.

The agritech company has helped in increasing yields by nearly 50% for about 140 farmers it works within the Nilgiri hills in Tamil Nadu. A precise day-to-day interventions to farmers is given by centrally managed Data Science and Recommendation Engines enabled by its data scientists and agronomists, which results in a 30 to 50% increase in yields. Aibono recently raised \$2 million in fresh capital from Rebright Partners and Mitsui Sumitomo Insurance Venture Capital taking its total funding to \$5 Million.

AgricxLab - Connects Cold Storage Owners to Agri-product Buyers

Agriculture Startup	AgricxLab
Founded In	2016
Headquarters	Thane
Founders	Ritesh Dhoot, Saurabh Kumar
Total Funding	\$500k
Investors	Ankur Capital, Centre for Innovation Incubation and Entrepreneurship

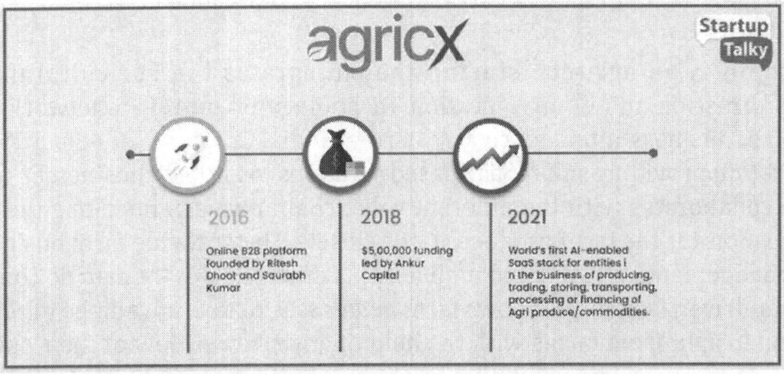

Agricx | Agribusiness Startups in India

AgricxLab is a Thane-based online B2B platform founded by Ritesh Dhoot and Saurabh Kumar for connecting cold storage owners with bulk buyers for agri-products. They use smartphone imaging to assess the quality of Agri-produce through its mobile app which uses artificial intelligence and computer vision to on images to yield objective, accurate and faster quality assessment of agri-produce. Agricx offers solutions to warehouses and enterprise clients with a plan to expand across the food produce supply chain. The startup has raised a total funding of $500K from various investors, including Ankur Capital.

Cropin - Provides Smart SaaS-based solutions to Agritech Businesses

Agriculture Startup	Cropin
Founded In	2010
Headquarters	Bengaluru
Founders	Krishna Kumar
Total Funding	$32.7 Million
Investors	Chiratae Ventures, Bill & Melinda Gates Foundation, ABC World Asia, Ankur Capital, BSP Funds and others

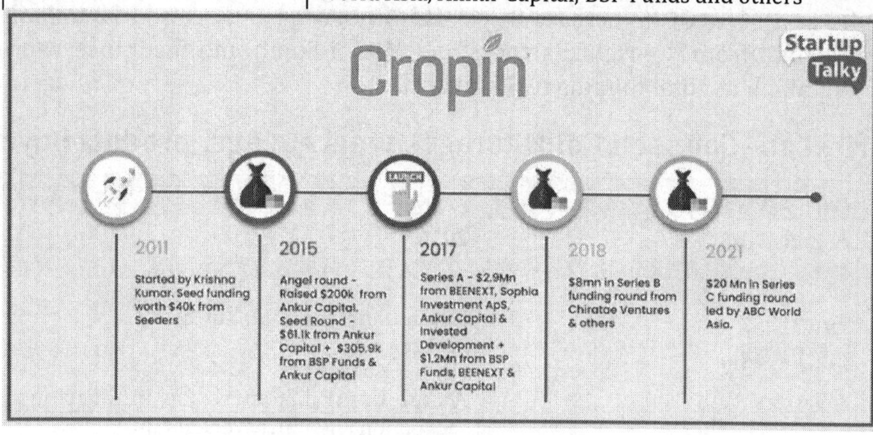

Cropin | Agro Startups in India

Cropin is an agritech startup that integrates the agricultural sector with Information and Communication Technology by putting a network of ERP and BI i.e. Business Intelligence across rural India. Cropin is Full-Stack Agritech organization providing smart SaaS based solutions to agritech businesses globally. Cropin collaborates with the different value chain participants along the supply chain to monitor the farm produce status closely. They provide farm businesses a farm management software and mobile app, which enables them to do connected and data-driven farming. It allows farm businesses to take advantage of real-time data and insight from farms with the help of an accurate view of their operation throughout the growing season and to improve financial, operational and agronomy aspects.

Cropin uses cutting-edge technologies, Big Data analytics, Artificial Intelligence, Geo-tagging & Satellite monitoring to revolutionize the agri-ecosystem. This agriculture startup has raised a total of $32.7 Million funding till date, with its recent Series C round in 2021 for $20 Million from ABC World Asia.

Fasal - Uses AI & Data Science to make On-Farm Predictions

Agriculture Startup	FASAL
Founded In	2018
Headquarters	Bangalore
Founders	Shailendra Tiwari and Ananda Verma
Investors	Mistletoe, Zeroth.AI, Mind Fund, Artesian VC

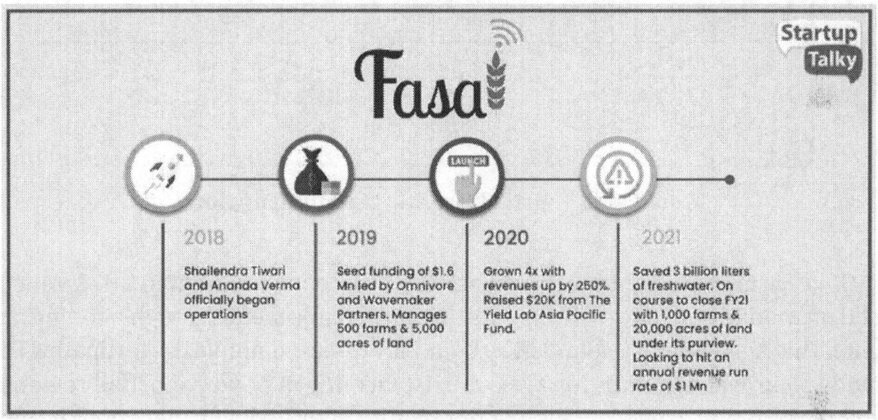

Fasal | Agribusiness Companies in India

FASAL is a Bengaluru-based agritech platform founded by Ananda Verma and Shailendra Tiwari in 2018 which is developed by Wolkus Technology Solutions and is an AI powered platform for agricultural ecosystem. It records a variety of growing conditions on the farm and then uses artificial intelligence and data science to make on-farm predictions, before delivering the insights anywhere on any device including Android, iOS, tablets and web.

FASAL has saved close to 3 Billion Liters of freshwater to date. This agritech startup is on a course to the close FY21 with 1000 farms and 20, 000 acres of land under its purview. It is also looking to hit and annual revenue run rate of $1 Million.

Gold Farm - Provides Agri Farm equipment via Booking Farm Agents

Agriculture Startup	Gold Farm
Founded In	2012
Headquarters	Bangalore
Founders	Abhilash Thirupathy
Investors	Mahindra Rise, Infuse Ventures, Intellecap

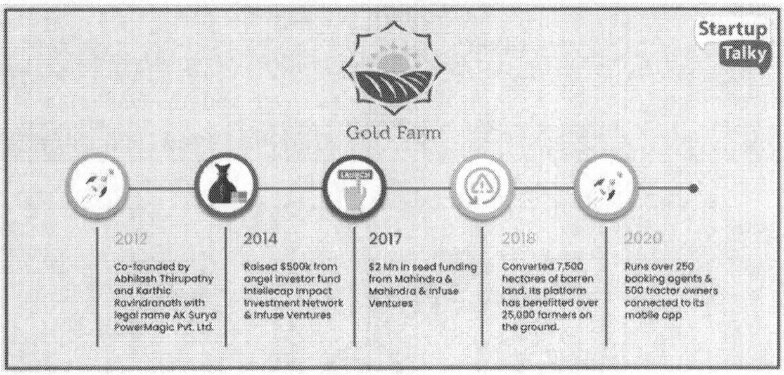

Gold Farm | Agriculture Startups in Bangalore

Gold Farm is a Bangalore-based agritech company founded in 2012 which is based on mobile app to provide agriculture farm equipment through booking farm agents. This Agri startup is founded by Coimbatore-based Abhilash Thirupathy. They provide solar water pumps for agriculturist farming in power shortfall regions in our country. Gold Farm provides an opportunity door to the farmers booking farm equipment through the call service and mobile application.

Currently, Gold Farm runs over 250 booking agents and 500 tractor owners connected to their mobile app. They aim at enhancing the growth and sustainability of farmers by structuring productive ecosystems through innovation and technology. The startup has raised a total funding of $3 million.

Ninjacart - India's Largest B2B Fresh Produce Supply Chain Company

Agriculture Startup	Ninjacart
Founded In	2015
Headquarters	Bangalore
Founders	Ashutosh Vikram, Kartheeswaran K K, Sharath Loganathan, Thiru Nagarajan, Thirukumaran Nagarajan, Vasu Devan

Total Funding	~ $194.2 Million
Investors	HR Capital, Walmart, Tiger Global Management, Trifecta Capital Partners, Flipkart, Nandan Nilekani & others

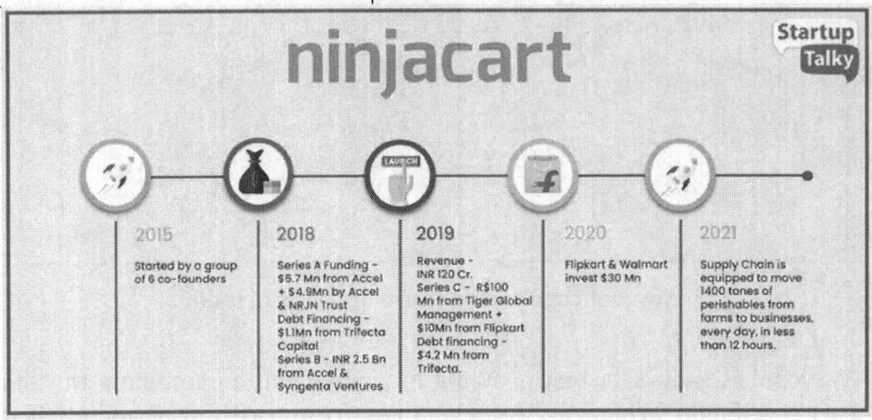

Ninjacart | Agriculture Supply Chain Companies in India

Ninjacart is just one of the few Bangalore-based agriculture supply chain companies in India. It was founded by Thirukumaran Nagaranjan in 2015. It currently moves more than 60 tons of produce a day from farm to store in less than 14 hours at a lower than traditional supply chains. Ninjacart helps more than 2,000 farmers to sell more than 80 vegetables and fruits every day to more than 800 retailers and restaurants in Bengaluru. They also claim to have wasted 4% wastage in the entire supply chain.

As of 2021, Ninjacart raised a total funding of around $194.2 Million of which Flipkart and Walmart recently invested $30 Million in this agritech supply chain startup. Currently, Ninjacart is equipped to move 1400 tons of perishable from farms to businesses, everyday, in less than 12 hours.

Waycool - Food Distribution Company

Agriculture Startup	Waycool
Founded In	2015
Headquarters	Chennai
Founders	Sanjay Dasari
Total Funding	$64.9 Million
Investors	LGT Impact Ventures, Caspian Impact Investments, Aspada, Northern Arc, InnoVen Capital, Lightbox & others

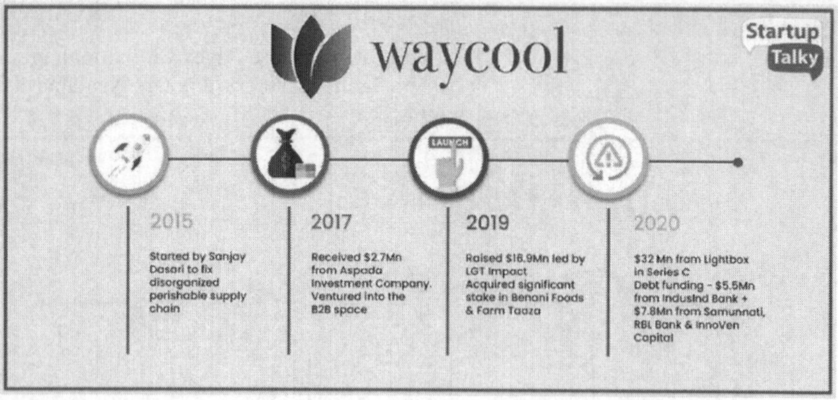

Waycool Logo | Agriculture Startups in India

Waycool is India's fastest growing food distribution company that has a network for 35,000 farmers across vegetables, fruits, rice, pulses and other food products. It was established in July 2015 with the aim of fixing the disorganized perishable supply chain. They have a retail presence, currently present in Chennai through the brand SunnyBee, operating retail stores, mobile stores on trucks, as well as e- commerce. They also have a large number of high-quality private label products that SunnyBee produces and brands through partnerships with sister companies.

This Agritech has raised a total funding of $64.9 Million, of which its recent funding came in 2021 from Lightbox Ventures ($32 Million - series C), IndusInd Bank ($5.5 million - debt) and $7.8 Million came from Samunnati, RBL Bank and InnoVen Capital. Waycool is actively using 19 technologies for its website and has $2 million estimated revenue annually.

Intello Labs - Leverages AI to assess Food Commodities' Quality

Agriculture Startup	Intello Labs
Founded In	2016
Headquarters	Gurgaon
Founders	Milan Sharma, Nishant Mishra, Himani Shah and Devendra Chandani
Total Funding	$7.9 Million
Investors	Indigram Labs, GenNext Venture, Nexus Venture Partners and Omnivore

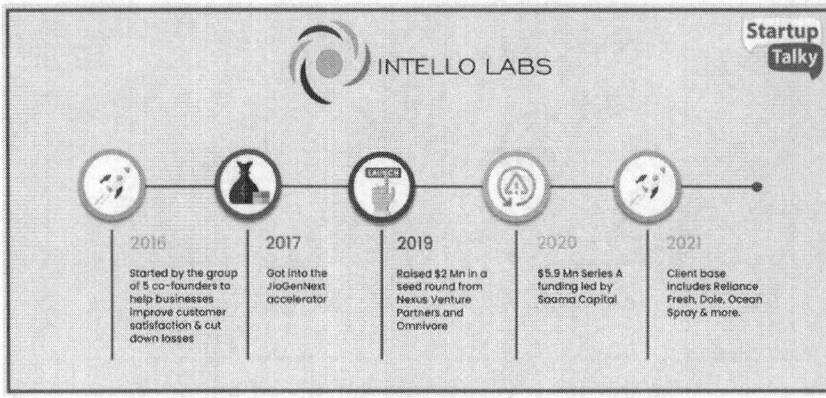

Intello Labs | Agritech AI Business in India

Intello Labs does quality assessment of food commodities using computer vision and AI. Intello Labs help food businesses like growers, traders, retailers, food service companies, exporters, etc. to improve their customer satisfaction and cut down losses.

Intello Labs reports to be aggressively growing in India while also pursuing global markets, including the likes of China, Southeast Asia, and the USA. Some of Intello Labs' larger clients include Reliance Fresh, Dole, Ocean Spray, and many of the leading e- grocery companies in China and Southeast Asia. In 2020, Intello Labs raised $5.9 million, which is approximately Rs 45 crore, in Series A funding led by Saama Capital.

Agrowave - Creating Mobile Pickup Stations for Sustainable Agri Supply

Agriculture Startup	Agrowave
Founder	Anu Meena
Founding Year	2017
Headquarters	Delhi
Investors	Shekhar Puli, Daffodil Software
Revenue	INR 27 Cr (FY20)

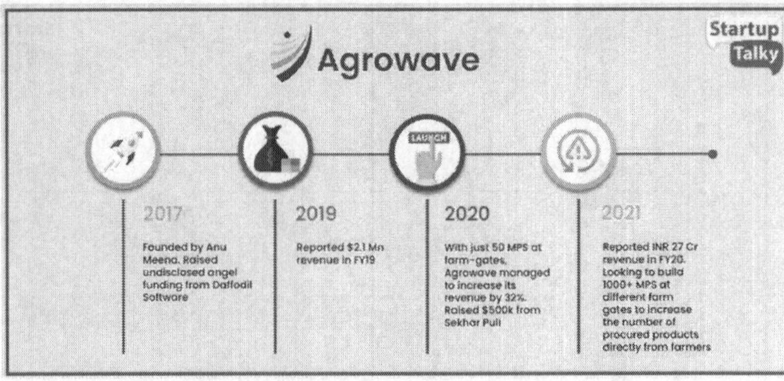

Agowave - Agriculture Startup in India

Founded by Anu Meena, Agrowave is an Agritech company that aims at helping farmers by directly connecting them to businesses and constructing a sustainable supply chain management. It has built several pick-up stations for farmers where they can load their products and that they don't have to travel very far to sell it. It has a mobile app running on IOs and Android, where farmers sell their produce, the delivery trucks pick up the produce from farmer gates and pick up stations and allow hassle-free payments. This produce is then delivered through systematic route mapping to the end customers.

Anu Meena, alumni of Delhi IIT, conceived this brainchild after witnessing the struggles and challenges of her grandfather back in Rajasthan who happens to be a farmer himself. She aims at simplifying business for farmers all over by eliminating middlemen and giving farmers the best price for their produce.

BharatAgri - Provides Personalized Advisory Service to Farmers

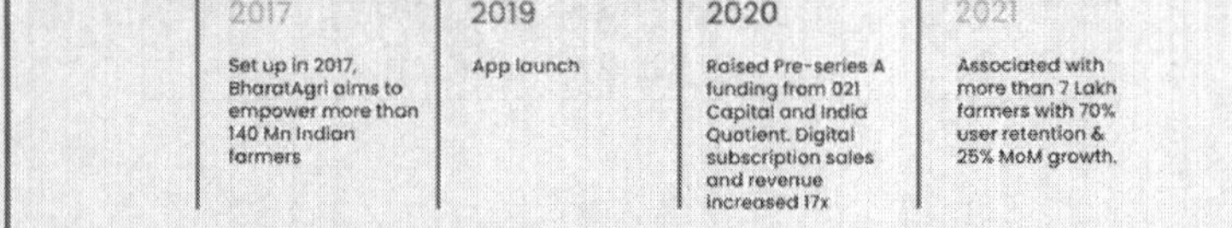

BharatAgri - Agriculture Startup in India

Agriculture Startup	BharatAgri
Founders	Sai Gole, Siddharth Dialani
Founding Year	2017
Headquarters	Pune
MoM Growth	25% as of 2021
Investors	021 Capital, India Quotient

BharatAgri is a personalized service where farmers are educated ab serout the crop they're planning to plant and provides complete guidance and expertise on how to grow it. The service includes expertise on pesticides, critical advisory on weather and soil analysis. It ensures maximum and yield on production with timely analysis of production costs. BharatAgri lends support at every stage of farming through offline and online technology including In-App chat support, Interactive voice response (IVR), and Whatsapp.

Co-founders Sai Gole and Siddharth Dialani quit their high-paying jobs in 2017 to learn about the challenges that farmers face and lived in a village near Pune to understand and implement their IIT learning experience into practice and find solutions to farmers' woes. With a farmer base of 3000 onboard, the agritech has now shifted to the B2F (Business to Farmer) model, so that they can provide better services to farmers needing technical help through digital farming.

BigHaat - Online Agri-input Distribution Platform

Agriculture Startup	BigHaat
Founders	Satish Nukala, Sachin Nandwana
Founding Year	2015
Headquarters	Bangalore
Investors	Ankur Capital, Beyond Next Ventures & more
Revenue	INR 80 Lakh (FY19)

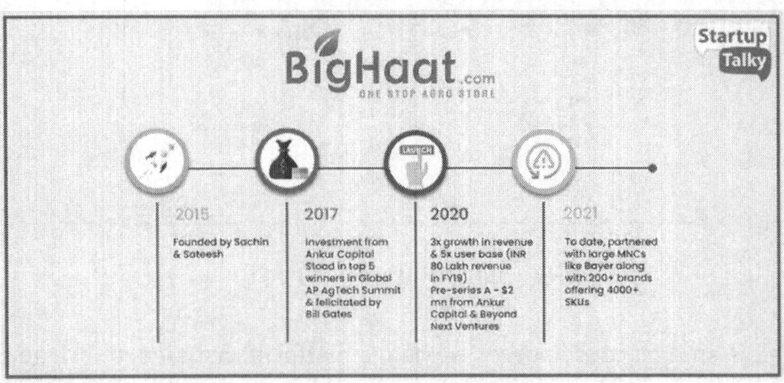

BigHaat - Agriculture Startup in India

BigHaat is an Agri input and digital marketplace for farmers who seek guidance and information on various aspects of farming. The agritech has a wide range of services and products such as advisory services related to crop nutrition and farming, products such as seeds, agricultural machinery, and much more.

It has also built an Agri platform in collaboration with Microsoft, called Kisan Vedika which addresses issues and queries through experts in regional languages. BigHaat's marketing team works both offline and online to acquire a customer base. In addition to this, they offer a support channel that is also available for any queries. The farmers have to give a missed call on the designated number and the Agri expert will revert with necessary guidance.

Founded in 2015 by Satish Nukala and Sachin Nandwana, BigHaat boasts about providing Agri advisor services and products to 4 million farmers in the country. The Agritech has raised 2 million in its latest funding series through investors such as Ankur Capital, Rockstud Capital, and Beyond Next Ventures.

Bijak - Bridges the gap between Agri Buyers and Sellers

Agriculture Startup	Bijak
Founders	Nikhil Tripathi, Mahesh Jakhotia, Nakul Upadhye, Jitender Bedwal, Daya Rai
Founding Year	2019
Headquarters	Gurugram
GMV	INR 2000 Crore (2020)
Investors	RTP Global, Sequoia, Omidyar
Total Funding	$14.5 Million

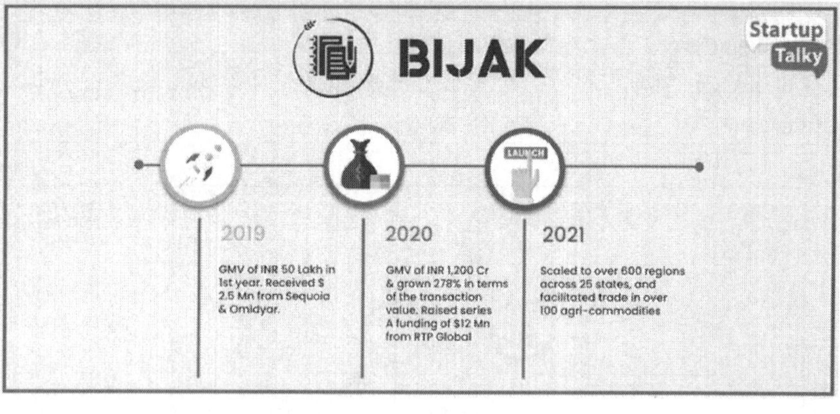

Bijak - Agriculture Startup in India

Bijak is an Agri commodities exchange platform that aims to bridge the gap between buyers and sellers. It also acts as a bookkeeping app that helps you keep all transactions and related documents on your phone. Bijak helps buyers with quick loan disbursals and immediate payments to suppliers, with real-time, hassle-free payments. It is spread over 27 cities and union territories and covers more than 100 agricultural commodities on its platform.

The B2B platform was founded in 2019 by a team of five namely Nikhil Tripathi, Mahesh Jakhotia, Nakul Upadhye, Jitender Bedwal, and Daya Rai. It has raised close to INR 91 crores in its Series A funding round. The ambitious startup aims at providing buyers and sellers a considerable working capital and improved logistics.

Clover Ventures - Offers Full-Stack Agronomy Solutions

Agriculture Startup	Clover Ventures
Founders	Arvind M, Avinash B R, Gururaj S Rao, Santosh Narasipura
Founding Year	2012
Headquarters	Bangalore
Total Funding	$7.5 Million
Investors	Omnivore, Accel, Mayfield & more

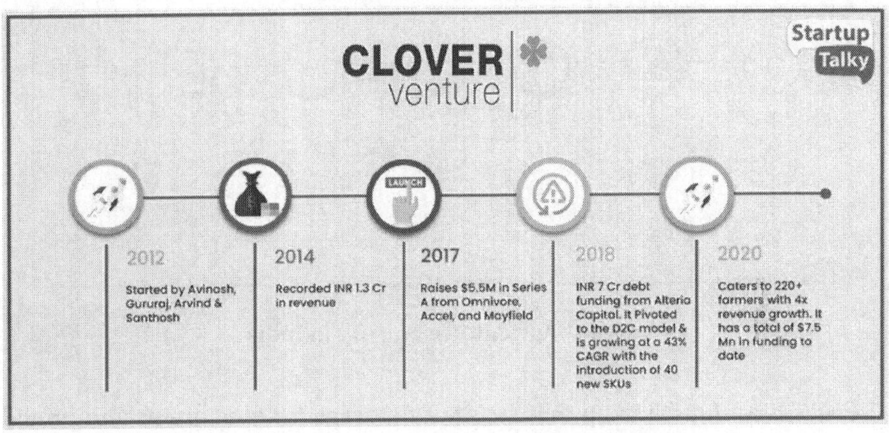

Clover Venture - Agriculture Startup in India

Clover Ventures is a Bangalore-based agritech company that deals in the creation and management of farm networks in Karnataka, Telangana, and Andhra Pradesh. It aims at tapping several agricultural roadblocks faced by farmers such as complex and inefficient supply chain, consistency and quality issues, and lack of stable backend supply. It aims at creating and solidifying an organized supply chain for fresh produce. The company currently manages over 70 acres of farmland and provides fresh produce to various fine dines, Retail stores, food processors, and online Internet Kitchens.

The logo with four leaves of the clover represents Arvind M, Avinash B R, Gururaj S Rao, and Santosh Narasipura, the co-founders of the company. The core functions of the company work on the model of the demand-backed supply chain for perishables. The key focus is on quality of the produce, consistency, and traceability. The B2B company is about to launch a B2C service out of Bangalore and Hyderabad this year.

Ergos - Building 'Grainbank' to Convert Food Grains into Digital Assets

Agriculture Startup	Ergos
Founders	Kishor Jha, Priyanka Kumari, Praveen Kumar
Founding Year	2012
Headquarters	Patna
Revenue	INR 19 Crore (FY19)
Investors	Aavishkaar Capital, Chiratae Ventures & more

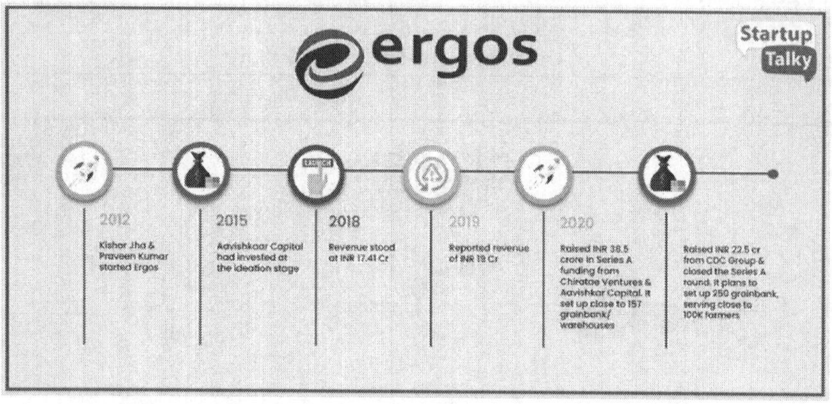

ergos - Agriculture Startup in India

Ergos is an Agri startup that facilitates storage for food grains and provides digitized solutions for the management of the same. Indian farmers who lack storage capacity often face grain loss in quantity and deterioration of quality and this leads to selling their produce at lower prices. Ergos helps them store their grains in their "Grain bank" and facilitates loans and supply chain management through networks with NBFCs and Banks. Farmers can always withdraw their produce, even a single bag of grains, at their disposal.

The Patna-based Agri startup was founded in 2012 by Kishor Jha, Priyanka Kumari, and Praveen Kumar. Ergos claims to support 25000 farmers so far and aims to extend its footprint outside Bihar by setting up 500-600 new grain Banks across various states. The company raised INR 81 crore in its recent funding round and is assertive on building better post-harvest and supply chain management.

FreshoKartz - Provides Soil Advisory and Fertilizers

Agriculture Startup	FreshoKartz
Founders	Rajendra Lora, Nagendra Yadav
Founding Year	2016
Headquarters	Jaipur
Revenue	INR 5.16 Crore (FY20)
Investors	Innovana ThinkLab, RVCF & more

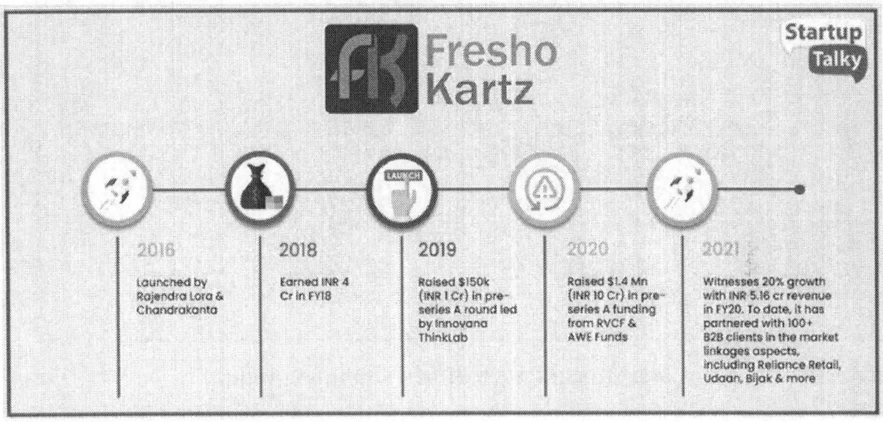

FreshoKartz - Agriculture Startup in India

FreshoKartz is an eCommerce marketplace for farm-fresh produce, based in Jaipur Rajasthan. It offers various services such as soil advisory and products such as pesticides, seeds, fertilizers, etc for farmers. The company works on the on-demand procurement of fresh farm produce and helps farmers avail themselves right prices. FreshoKartz offers transparent weighing and immediate payment of the produce. It also offers financing facilities for farmers during the harvesting season.

Founded in 2016 by Rajendra Lora and Nagendra Yadav, FreshoKartz is funded by iStart, an incubator run by the Rajasthan Government to support startups. Recognized by the government early on, the company has been provided with a co-working space and best knowledge imparting mentors in the Agri tech industry. These mentors help the company find VCs and angel investors. The company has also been awarded the Bamasaha techno fund worth INR 20 lakhs and a marketing fund worth INR 10 lakhs. FreshoKartz aims at eliminating middlemen in the process of selling fresh produce directly from the farmers to other businesses. Their B2B clientele so far includes ITC hotels, Hilton Jaipur, and several other commercial kitchens in Jaipur.

MeraKisan - Offers Certified Organic Products to Consumers

Agriculture Startup	MeraKisan
Founder	Prashant Patil
Founding Year	2014
Revenue	INR 30 Crore (FY20)
Investors	Mahindra Univeg, Ajinkya Rahane

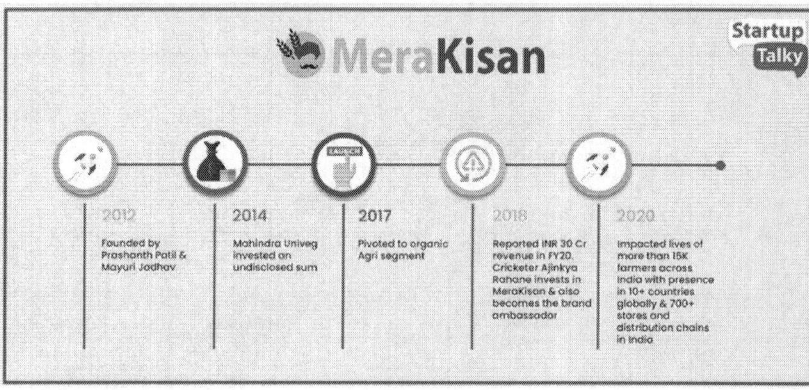

MeraKisan - Agriculture Startup in India

MeraKisan is an online marketplace for a wide range of freshly produced farm products. The farm to fork company aims to support local farmers by providing seasonal and regional fruits and vegetables directly to consumers. Its products range from pulses and cereals to exotic and seasonal vegetables along with superfoods and organic cooking oils. The company recently launched its own range of handcrafted juices, smoothies, salads, and refreshments under the brand Saboro Lounge. It also provides end-to-end soil advisory and other farming facilities.

MeraKisan was roped in by a joint venture between Mahindra Univeg Pvt. Ltd and Belgium-based Univeg (Greenyard Foods). Prashant Patil, the founder of MeraKisan is an IT engineer by profession and started operating in 2014. Patil, through a fresh issue of shares, holds 66% of MeraKisan Pvt. Ltd. The company claims to have more than 500 sellers on its website. In March 2020, the Mahindra-backed company found a new investor in Ajinkya Rahane, the vice-captain of the Indian Test Cricket team.

Annexure-IV

Entrepreneurial Test

Know your entrepreneur within you

Have you got the skills to think of an idea like, Ninjakart, Amul, ITC, Facebook, Yahoo ,Google &Twitter and turn it into a Profitable Business?

1:Totally disagree, 2: Somewhat disagree, 3: Somewhat agree, 4: Totally agree.

	1	*2*	*3*	*4*
1. I like to give myself challenges when I take on a new project				
2. I am fairly at ease in difficult situations				
3. Where others see problems, I see possibilities				
4. I always worry about what others will think before doing something important				
5. I am fairly curious and I am continually in search of discovery				
6. I am a lot less effective in stressful situations				
7. I want to build something that will be recognized publicly				
8. When faced with difficulties, I look for alternative solutions				
9. For me, what counts is action				
10. For me, it is possible to influence one's destiny				
11. I am capable of imagining how we can make things work				
12. When I take on a project I have confidence that I will carry it out successfully				
13. I shoot for excellence in everything I do				
14. According to me, we somehow make our own luck				
15. In general, I distrust my instincts				
16. I have no problem working for someone else				

	1	2	3	4
17. I try to be the first or the best in my area of competency				
18. For me, taking risks is like buying a lottery ticket: it's a question of chance				
19. I am capable of seeing many solutions to a problem				
20. I prefer having the final say				
21. I always try to learn lessons from my failures				
22. For me, everything is possible if I believe I can do it				
23. I prefer using the good old ways of doing things				
24. Success is mostly luck				
25. A certain level of stress stimulates me				
26. It is easy for me to motivate others to work with me				
27. I often feel stuck by a difficult situation				
28. I can easily imagine many ways to satisfy a need				
29. After a failure, I am able to pick myself up and start over				
30. I am not always ready to make sacrifices in order to succeed				
31. I don't like to influence others				
32. To be satisfied with myself, I take on easy projects				
33. When I take on a project, I am not always convinced that I can carry it out successfully				
34. I like to lead others				
35. I always try to take calculated risks				
36. I have a hard time functioning in uncertain or ambiguous situations				
37. I am always in the midst of launching new projects				
38. I have a hard time anticipating events, trends				
39. I really enjoy situations where there are rules to respect				
40. Today, without a lot of money, we cannot take on a whole lot				
41. Being too ambitious is often perceived poorly				
42. I have a tendency to put off difficult tasks until later				
43. I am the kind of person to see the glass as half empty instead of half full				
44. There is a time for thought but action is more important				
45. I am not afraid to take on initiatives				
46. No matter what we do, it doesn't depend on us				
47. I prefer being my own boss				

	1	*2*	*3*	*4*
48. I don't consider myself more ambitious than others				
49. I always give the best of myself in everything I do				
50. I manage my stress well in ambiguous and uncertain situations				

Annexure-V

Start-up Registration India

Who can register with start-up India?

An entity incorporated as a Private Limited Company, Partnership Firm or a Limited Liability Partnership can register themselves under the start-up India scheme. The annual turnover of these business entities should not exceed Rs.100 crores, and they should have been in existence for up to ten years from the date of its incorporation/ registration. Such an entity should be working towards innovation, development or improvement of products or services or processes.

What are the benefits of signing up with start-up India?

There are a number of benefits start-ups receive under the Start-up India Initiative. Nevertheless, in order to avail of these benefits, an entity is needed to be recognised by the DPIIT as a start-up.

Start-ups are allowed to self-certify their compliance with six labour laws and three environment laws. This is allowed for a total period of five years from the date of incorporation/registration of the entity. Start-ups are allowed a three-year tax exemption and the best intellectual property services and resources solely built to help start-ups protect and commercialise their IPRs.

What kind of business structure should I choose for my start-up?

The most preferred business structures for a start-up are Private Limited companies and LLPs. A Private Limited company is legally recognized and generally favoured by investors. However, it has stricter compliance and may have a higher cost of incorporation.

Whereas incorporation cost is lower for LLPs and they tend to have relaxed compliance in comparison to the Private Limited Companies. In addition to that, LLPs have limited liabilities and are equally recognised by investors and all over the world.

What can I do to attract investors to a start-up?

To attract investors, not only do you need a stellar product with a scalable model, but you also need visibility. Make sure that your product receives healthy engagement and traction. You'll need to register your start-up-on-start-up India and proactively seek out investors. Make sure you are able to effectively communicate your business idea to the investor and the sustainability of your business model.

Can a foreign company register under the Start-up India hub?

Any entity that has at least one registered office in India can register itself on the hub, since the location preferences, for the time being, are only created for Indian states. However, soon the government hopes to start registrations for stakeholders from the global ecosystem too.

What is the difference between an accelerator and an incubator?

Start-up incubators are typically institutions that help entrepreneurs by developing their business, especially in the initial stages. The incubation function is usually carried out by institutions that have experience in the business and the tech world.

Start-up accelerators support early-stage, growth-driven companies. These programmes usually have a timeframe in which individual companies spend anywhere between a few weeks and a few months working with a group of mentors who are educated and may also provide financial help.

For how long is a company recognised as a start-up?

Any business entity that has completed 10 years from the date of its incorporation/registration, and has exceeded the previous year's turnover of Rs.100 crores shall stop being a start up on completion of 10 years from the date of its registration/incorporation.

Can an existing entity register itself as a "Start-up" on the Start-up India Portal?

Yes, as per the law an existing entity can register itself as a start-up, provided that it meets the prescribed criteria for a start-up. They will also be able to avail various tax and IPR benefits that are available to start-ups. The criteria are the same as those mentioned in the article above.

How do I know my registration is complete?

Once the application is complete, and the start-up gets recognised, you will receive a system-generated certificate of recognition. You will be able to download this certificate from the Start-up India portal.

Benefits under Start-up India

Simple process

The government of India has launched a mobile app and a website for easy registration for start-ups. Anyone interested in setting up a start-up can fill up a simple form on the website and upload certain documents. The entire process is completely online.

Reduction in cost

The government also provides lists of facilitators of patents and trademarks. They will provide high-quality Intellectual Property Right Services including fast examination of patents at lower fees. The government will bear all facilitator fees and the start-up will bear only the statutory fees. They will enjoy 80% reduction in the cost of filing patents.

Easy access to Funds

A 10,000 crore rupees fund is set-up by government to provide funds to the start-ups as venture capital. The government is also giving guarantee to the lenders to encourage banks and other financial institutions for providing venture capital.

Tax holiday for 3 Years

Start-ups will be exempted from income tax for 3 years provided they get a certification from Inter-Ministerial Board (IMB).

Apply for tenders

Start-ups can apply for government tenders. They are exempted from the "prior experience/turnover" criteria applicable for normal companies answering to government tenders.

R & D facilities

Seven new Research Parks will be set up to provide facilities to start-ups in the R&D sector

No time-consuming compliances

Various compliances have been simplified for start-ups to save time and money. Start-ups shall be allowed to self-certify compliance (through the Start-up mobile app) with 9 labour and 3 environment laws.

Tax saving for investors

People investing their capital gains in the venture funds setup by the government will get exemption from capital gains. This will help start-ups to attract more investors.

Choose your investor

After this plan, the start-ups will have an option to choose between the VCs, giving them the liberty to choose their investors.

Easy exit

In case of exit – A start up can close its business within 90 days from the date of application of winding up

Meet other entrepreneurs

The government has proposed to hold Two start-up fests annually both nationally and internationally to enable the various stakeholders of a start up to meet. This will provide huge networking opportunities.

Start-ups are being highly encouraged by the government. The benefits enjoyed by them are immense, which is why more people are setting up start-ups.

Annexure-VI

List of Leading Incubators for grant and incubational support in India

Incubators are the organizations that work to promote innovation and entrepreneurship among citizens. They basically provide funds and their infrastructure to the startups so that they can carry out their business activities of development, product trial, market entry, commercialization, etc. The government provides incentives to the incubators and the incubators will further provide funds to the startups. Under the Startup India Seed Fund Scheme 2022, the government is going to provide a grant to 300 incubators. The funds provided to the startup under this scheme will be up to Rs 50 lakh. The incubators can apply under the startup India seed fund scheme through the official portal. After verifying their application, the government will grant them the seed fund. The startups can also apply under this scheme directly from the portal and from there they can select the incubators of their choice according to their preference.

S.No.	Name	Thrust Area	State	Website Application Process	Contact Details
1	NASSCOM 10K Warehouse Vizag	Agnostic	Andhra Pradesh	http://10000startups.com/startup-warehouse/	Vijay Kumar Bawra, Manager vijay@nasscom.in M: 9831524485
2	Agri Business Incubator	Agribusiness	Andhra Pradesh	http://www.aipicrisat.org/	Dr Kiran Sharma, CEO k.sharma@cgiar T:: 040 3071 3300 /3417
4	IKP Knowledge Park Life Science Incubator	1) Life Science, 2) Medical Devices, 3) Materials	Andhra Pradesh	http://www.ikpknowledgepark.com/index.php	Dr. Sangita Sen Majee, Head – Life Science Incubator sangita@ikpknowledgepark.com Fax: 040 3071 3074/75
5	IITG-Technology Incubation Centre (IITGTIC)	1) IT 2) Healthcare 3) Renewable enrgy 4) Mechanical	Assam	http://www.iitg.ac.in	Dr. J. K. Deka, Faculty In-Charge tic@iitg.ernet.in T: (0361) 2583191/2583194 Fax: 0361-2583195
6	Centre for Innovation Incubation and Entrepreneurship	Agnostic	Assam	https://adtu.in/	Mr. Joutishman Dutta, Principal jd@downtowngroup.org M: 9706011569
7	Bihar Entrepreneurs Association	Agnostic	Bihar	http://www.enterprisingzone.com	Mr. Abhishek Kumar, BEA abhishek2709@gmail.com M: 09708899777 T: 0612-3222433
8	Bihar Industries Association	Agnostic	Bihar	http://www.biabihar.com/	Mr. Ram Lal Khaitan, BIA, hi.techpatna@gmail.com M: 09334145197 T: 0612-2226642/2222100/3260717
9	Business Incubation Centre (IC)	1) Electronics System Design and manufacturing (ESDM) with a special focus on Medical Electronics	Bihar	http://www.iciitp.com/	Aditya Nataraja, Manager sriru@iitp.ac.in T: +91-612-3028545/6/7
10	Foundation for Innovation and Technology Transfer, IIT Delhi	Science & Technology	Delhi	http://fitt-iitd.in/	Dr. Anil Wali, Managing Director 1) anilwali@fitt.iitd.ac.in 2) mdfitt@gmail.com T: 011-26857762

S.No.	Name	Thrust Area	State	Website Application Process	Contact Details
11	National Small Industries Corporation Limited	1) Energy 2) Pumps	Delhi	http://www.nsic.co.in/	Mr.Satvinder Singh, Senior General Manager 1) pkbiswas@gmail.com 2) gmtech@nsic.in T:#-011-26317482, Fax:011-26317482
12	Shriram Institute for Industrial Research (A Unit of Shriram Scientific and Industrial Research Foundation) (www. shriraminst itute.org)	Multidimensional services in the manufacturing sector of Plastics, Rubbers, Specialty Chemicals and Waste Management	Delhi	http://www. shriraminstitute.org/ technology-business-incubator.html	1) Dr. Bhupesh Sharma,Manager, Shriram Institute 2) Dr. Ajay K. Tyagi, Asst. Director 1)sritbi@ shriraminstitute.org 2)aktyagi@ shriraminstitute.org T: 91-11-27667267, 27667860, 27667983 Fax: 91-11-27667676, 91-11- 27667207 M: 9871438033
13	Atma Ram Sanatan Dharma College University of Delhi	1) Nanomaterials and Composite Technologies 2) Renewable energy applications 3) electronics and ICT related 4) energy /technologies 5) Waste to energy technology 6) e- waste 7) mobile computing 8) Embedded systems 9) Digital image processing 10) Web technology 11) Power System and Power Electronics 12) foods and dairy products 13) bioinformatics 14) cheminformatics 15) research and technology development 16) material sciences	Delhi	http://arsdstar.wixsite. com/ciel	Dr. Rajeev Singh, Incubator Manager 1)principal.arsdcollege@gmail.com 2) rajeev@arsd.du.ac.in M: 9311225501
14	Electropreneur Park by STPIDelhi/Delhi University/IESA	Electronics	Delhi	http:// electropreneurpark. res.in/	Mr. Himanshu Shekhar, Manager Admin himanshu@electropreneurpark.co m M: 91-8860940000 T: 011-65164488
15	M/s Zonal Technology Management - Business Planning and Development Unit" - ZTM-BPD, IARI, New Delhi	Agriculture & allied sectors	Delhi	http://ztmbpd.iari.res. in/?q=content/arise	Dr. Neeru Bhooshan, Incharge 1) zonaltech@iari.res.in 2) ps.ztmiari@ gmail.com T: +91-11-25843542, +91-11- 25843553

S.No.	Name	Thrust Area	State	Website Application Process	Contact Details
16	Centre for Incubation and Business Acceleration	1. ICT 2. Food Processing 3. Clean Technologies	Goa	http://www.ciba.org.in	D.S. Prashant General Manager prashant@ciba.org.in M: 9823235890
17	CIIIE Initiatives, Centre for Innovation, Incubation and Entrepreneurship (CIIE)	Market oriented products/technologies (ICT, Renewable energy, Social impact)	Gujarat	http://www.ciie.co/	Mr. Neeraj Jain, Manager neeraj@iima.ac.in T: 079 6632 4203 Fax: 079-66324207
18	Gujrat Foundation for Entrepreneurial excellence- icreate	Technology & innovation	Gujarat	http://icreate.org.in/	Mr. Anupam Lavania , Manager anupam@icreate.org.in M: 079-27912803
19	IITGN Incubation Centre	Technology	Gujarat	https://www.iitgn.ac.in/sites/ default/files/IIEC-Brochure-2016.pdf	Mr. Anand Pandey anandp@iitgn.ac.in M: 07069021616
20	Comcubator (www.micaedc.or g)	Communication technology	Gujarat	https://www.mica.ac.in/edc/ mica-incubator	Prof U T Rao, Chairperson rao@micamail.in T: 02717 308319 Fax:02717 308249 M: 9979777284
21	Gujarat State Biotechnology Mission (GSBTM), Savli Bio Incubator	Biotechnology	Gujarat	http://www.savlibioincubator.in/	1) Shri. S.N.Tyagi, Mission Director & Project Director 2) Dr. A. N. Bhadalkar,Sector Specialist & Project co-coordinator 3) Mrs. Anasuya Bhadalkar, Sector Specialist & Project co-coordinator 1) mdbtm@gujarat.gov.in 2) ssibtm@gujarat.gov.in 3) gujaratbiotech@gmail.com 4) info@savlibioincubator.in T: 079-23252197, 079-232 52164, 2667266000
22	B. V. Patel Pharmaceutical Education & Research Development (PERD) Centre	Biotechnology	Gujarat	http://perdcentre.com/BIRAC.html	Dr. Neeta Shrivastava, Assistant Director 1) perd@perdcentre.com 2) neetas@perdcentre.com T: 079 27439375/2741640

S.No.	Name	Thrust Area	State	Website Application Process	Contact Details
23	Ahmedabad University Support Foundation, Venture Studio, Ahmedabad University	1) Healthcare 2) Education 3) Energy 4) Transport 5) Agritech 6) Hardware & Manufacturing	Gujarat	http://www.venturestudio.in/	Ms. Shanti Menon, Management Administration vs@ahduni.edu.in T: 079 2646 3150
24	Innovation & Incubation Centre (IIC) Pandit Deendayal Petroleum University(PDPU)	1) Energy and Aligned areas 2) IT 3) E commerce	Gujarat	http://www.iic.pdpu.ac.in/voca.html	Shani Panyda, Executive shani.pandya@iic.pdpu.ac.in T: 9409611194 / 07923275163
25	Society for Innovation & Entrepreurship in Dairying (SINED) Technology Business Incubator	Dairy technology	Haryana	http://www.ndritbi.com/	Dr. Ashish Kumar Singh, Principal Scientist & Secretary SINED-TBI aksndri@gmail.com T: 0184-2259304, 2259329
26	91springboard, Gurgaon	1) Sector neutral (vertical) 2) Internet & Mobile Technology (Horizontal)	Haryana	www.91springboard.com	Pranay Gupta, Cofounder info@91springboard.com +91-9015419191
27	Shri Mata Vaishno Devi University Technology Business Incubation Center Society (SMVDU TBI)	1) Biotechnology (Microbial, Plant, Medical Genetics and Diagnostics) 2) Engineering (Electronics, Robotics, E-commerce, GSM/GPRS based Technology, Alternative Energy Management) 3) Business Development	Jammu & Kashmir	http://smvdutbic.org/tbic/	Dr Swarkar Sharma, Interim CEO smvdu.tbic@smvdu.ac.in T: 01991-285535 // 285524 // 285634 // 285699 extn: 2385
28	Indo Danish Tool Room	Mechanical	Jharkhand		Mr.Ashutosh Kumar, Dy. General Manager reach@idtrjamshedpur.com T: 0657-2201261/62 Fax: 0657-2202723
29	NASSCOM-ERNET Center Of Excellence (CoE) For Internet Of Things (IoT)	Agriculture, Healthcare, Energy, Transportation, Emerging Technologies	Karnataka		Sanjeev Malhotra, CEO Coe-iot@nasscom.in 080 41693924

S.No.	Name	Thrust Area	State	Website Application Process	Contact Details
30	Composites Technology Park (www.composites technologypark.c om)	Composites based on coir, bamboo, jute	Karnataka	www.compositestech-nologypark.com	Dr. R Gopalan, Executive Director drgopalan2003@yahoo.com T:[illegible]-[illegible] / [illegible] / 28482768 Fax: 080- 28482771
31	Syndicate	1) Blockchain 2) Internet of things 3) Fintech 4)Logistic Technology	Karnataka	http://www.thesyndicate.tech /	Rammohan, rammohan.prabhakar@thesyndicat e.tech M:9845288752
32	IIITB Innovation Center	IT	Karnataka	http://www.iiitb.org/	Mr. D.V.Jagadish, CEO 1)startup@iiitb.org 2)dv.jagadish@iiitb.ac.in T: +91 80 4140 7777 Fax: +91 80 4140 7704
33	MSME Business Incubator.	1) IT 2) Biotech	Karnataka	http://nmamit.nitte.edu.in/	Dr. Balasubramani R., Chief Project Leader - EDC 1)balasubramni.r@nitte.edu.in 2)info@nitte.edu.in Fax: 91-0824-2204305,08258- 281265 M: 9900462705
34	Science & Technology Entrepreneurs Park (BECSTEP)	1) Food Processing 2) Textiles 3) Building Technology	Karnataka	http://www.becbgk.edu/step. php	Dr Bharti S. , Executive Director M: 9448972628
35	STEP-SJCE	1) Electronics & IT 2) Automotive design 3) Mobile applications 4) Data analytics 5) Embedded systems 6) Energy conservation 7) E-marketing	Karnataka	http://www.sjcestep.in/	Dr. B G Sangamashwera, Chief Executive info@sjcestep.in T: 0821-2548321 Fax: 0821-2548321
36	Global Incubation Services (GINSERVE) (Web: www.ginserve.in)	Technology	Karnataka	http://www.ginserv.in/	Mr. Vinod Shankar, COO vinod@ginserv.in M: 984575103
37	Nadathur S Raghavan Centre for Entrepreneurial Learning (NSRCEL)	Agnostic	Karnataka	http://www.nsrcel.org/	Rajiv Sawhney, COO rajiv.sawhney@IIMB.ERNET.IN M: +91 8026993758
38	C-CAMP Bioincubator	1) Life Sciences 2) Biotechnology	Karnataka	www.ccamp.res.in	Dr. Taslimarif Saiyed, Director and COO(C-CAMP) taslim@ccamp.res.in T: 91-80-6718-5100

S.No.	Name	Thrust Area	State	Website Application Process	Contact Details
39	National Design Business Incubator (NDBI)	Design	Karnataka	www.ndbiindia.org	Srinath Bhadram Manager manager.ndbi.b@nid.edu M: 9902067555 T: 080-23408200
40	Technovate Innovations	1) Healthcare 2) Logistics	Karnataka	http://venturefactory.in	Mr. Vincent Rajendran, Director Operations v.rajendran@i2india.in M: 9535724747
41	LetsVenture (Indiepitch Solutions Pvt. Ltd.)	1) Technology 2) IT/ITES 3) Software 4) Hardware 5) Sector Agnostic	Karnataka	www.letsventure.com	Atit Danak, Director atit.danak@letsventure.com +91-9910491315
42	Excubator Consulting Private Limited	1) IoT 2) Artificial Intelligence 3) IT +C110	Karnataka	www.excubator.org	Rakesh Mishra, Director & cofounder rakesh.mishra@excubator.org 9845032742
43	TLabs-Times Internet Limited	1) Technology 2) E-Commerce 3) Online Media 4) EdTech 5) Healthcare 6) Artificial Intelligence 7) Machine Learning 8) Internet of Things 9) AdTech 10) FinTech	Karnataka	http://tlabs.in/	Abhishek Mitra Gupta, Chief Operating Officer ad'hishek@tlad's.iŶ +[illegible]
44	Amal-Jyothi Rural Technologies Business Incubator	1) Rural Technology 2) Green Technology 3) Information Technology	Kerala	http://www.ajce.in/tbihome/	Mr. Jose Kannampuzha, Principal principal@amaljyothi.ac.in T: + 91 4828 305503, 305626 M: 9496805712
45	Startup Village	Agnostic	Kerala	https://www.sv.co/	Mr. Sanjay Vijayakumar, Chairmain and Managing Director sanjay@sv.co M: 9846819123
46	CET - Technology Business Incubator	1) Green Technology 2) IT	Kerala	http://www.cet.ac.in/	Dr. A Samson , Coordinator asamson64@gmail.com T: 0471-2515502, 0471-2515556 M: 9447324844
47	KSIDC, Trivendrum	Biotech	Kerala	http://www.ksidc.org/	Biju B G , Assistant General Manager bijubg@ksidcmail.org M: 09847936409 T: 0471-2318922
48	NASSCOM 10000 Startup Warehouse Kochi	1) IT Products 2) IOT 3) SAAS	Kerala	http://10000startups.com/startup-warehouse/	Arun Nair, Deputy Manager arun@nasscom.in T: 0484 2415140/50

S.No.	Name	Thrust Area	State	Website Application Process	Contact Details
49	TBI (http://nitc.ac.in/nitc/misc/tbi/pub lic_html/index.ht m)	1) IT & electronics 2) Agriculture Engineering 3) Nano-tech	Kerala	http://www.tbi.nitc.ac.in/	Dr. Abraham T Mathew, Chairman, TBI (Dean -Research & Consultancy) NITC atm@nitc.ac.in T: 0495-2286144 M: 9447087949 FAX: 0495-2287250
50	Technopark TBI	1) IT 2) Electronics 3) Biotechnology 4) E-learning 5) Bio informatics 6) Social Entrepreneurship	Kerala	http://www.technopark.org/	Dr. Jayasankar Prasad , CEO, T-TBI ceo@startupmission.in T:: 471- 2700270 Fax: 471- 2700171
51	Amrita TBI	IT	Kerala	http://amritatbi.com/	Snehal Shetty, Vice President snehal@amritatbi.com T: 0476 280 5454
52	Technology Business Incubator for Medical Devices & Biomaterials (SCTIMST-TIMed)	1) Medical devices 2) Biomaterials 3) Healthcare	Kerala	http://timed.org.in/	Dr. S. BALRAM, CEO timed@sctimst.ac.in M: 9249402311
53	KSIDC Business Incubation Centre (KSIDC-BIC)	IT & ITES	Kerala	http://www.ksidc.org	Shri. R.Prasanth- Deputy General Manager prasanth@ksidcmail.org + 91-9847340444
54	Technology Innovation and Incubation Centre (TIIC)	ITES & Products	Madhya Pradesh	http://www.tiiciiitm.com/	Dr.Anurag Srivastava, Coordinator profanurag@gmail.com T:(O) 0751-2449826 M: +91-9826189049
55	Venture Center	Agnostic	Maharashtra	http://www.venturecenter.co.	Dr.V.Premnath, Director 1) v.premnath@ncl.res.in 2)director@venturecenter.co.in ,020-2590-2185 Fax: 91- 20-2589-3104 Phone: 020-2590-2986, 020-6401-1024
61	NASSCOM 10000 Start Ups Warehouse – Navi Mumbai	1) IT and enabled Technologies	Maharashtra	http://10000startups.com/10k-program/	Sahil Gupta, Deputy Manager sahil@nasscom.in T: 022-27784414
62	UnLtd India	Social Startups	Maharashtra	http://www.unltdindia.org/	

S.No.	Name	Thrust Area	State	Website Application Process	Contact Details
63	MITCON Technology Business Incubation Centre (A Division of MITCON) Consultancy & Engg. Services Ltd.)	1) Biotechnology 2) Ayurveda 3) Biomedicine 4) Renewable Energy 5) Environment 6) IT 7) Engineering electronic sectors	Maharashtra	https://www.mitconbiopharma.com	Dr. Pradeep Bavadekar, Managing Director md@mitconindia.com T: 020 - 66289452 M: 08308809265
64	Science And Technology Park	1) Open Source & Open Platform Technology 2) Renewable Energy & Clean Technology 3) Pharma & Biotechnology 4) Mobile Computing 5) Project Management 6) Data Centers 7) Social Incubation 8) IT/ITES 9) Education 10) Agri & Food Processing Technology 11) Remote Sensing & GIS 12) Electronics & Telecommunications 13) Cyber Security 14) Health	Maharashtra	http://scitechpark.org.in/	Dr.Rajendra P.Jagdale, Director General stp@scitechpark.org.in T: +91-20-25699206 / 25693449
65	SINE (Society for Innovation &Entrepreneurship)	Agnostic	Maharashtra	http://sineiitb.org/sine/home /	Mr. Rakesh Rajiv , Business Executive 1) sine@sineiitb.org 2) rakesh@sineiitb.org T: +91 22 2576 7016 Fax: +91 22 2572 1220
66	DKTE Technology Business Incubator	Clothing and textile technology	Maharashtra	http://www.dktetbi.com/	Prof.(Dr.) P.V.Kadole, Principal pvkadole@gmail.com T: 0230-2421300
67	NASSCOM 10000 Startup Warehouse Pune	1) IT 2) Software	Maharashtra	www.10000startups.com	Darryl Zuzarte Manager darryl@nasscom.in M: 9850813443
68	Trident Business Incubator	1) Biotech 2) Internet Startup 3) Engineering Industry	Odisha		Mr Santosh Kumar Sahoo, Assistant Professor 1)edcell@tat.ac.in 2) esantoshsahoo@gmail.com M: 9861387561
69	KIIT - Technology Business Incubator (KIITTBI)	1) Functional food 2) Bioinformatics 3) Exploitive Micro-biology 4) Agri-tech & Information Technology 5) Electronics and engineering products	Odisha	http://www.kiitincubator.in/	Dr. Mrutyunjay Suar, CEO mrutyunjay@kittincubator.in M: 9437011465

S.No.	Name	Thrust Area	State	Website Application Process	Contact Details
70	Guru Nanak Dev Engineering College STEP	1) Mechanical 2) IT	Punjab	http://stepgndec.com/	Prof. Nirmal Singh , Executive Director contact@stepgndec.in step.gne@gmail.com T: 0161-2814748 M:9855020485
71	Science and Technology Entrepreneur's Park (STEP)	1) Biotechnology 2) IT 3) Electronics 4) Mechanical	Punjab	http://www.thapar.edu	Dr. Dinesh Goyal, Executive Director dgoyal@thapar.edu M: +91 9815601705
72	Startup Oasis	Agnostic	Rajasthan	http://startupoasis.in/	Mr. Dilnawaz Khan, Program Manager connect@startupoasis.in M: 9887027865
73	BITS Pilani- Technology Business Incubator	Embdedded systems & VLSI designs	Rajasthan	http://www.bits-pilani.ac.in/pilani/technologybusiness/TechnologyBusinessIncubator	Dr. Anu Gupta, Program Manager anug@pilani.bits-pilani.ac.in T: +91-1596-51-5694, 9828125263 Fax: 91 01596 244183
74	A.C.Tech Campus	Biotech Engineering	Tamil Nadu	https://www.annauniv.edu /	Dr.B.A.Muralidhar, Assistant Director murlidharba@annauniversity.edu M: 9444539875
75	Adhiyamaan College of Engineering - TBI	1) GPS 2) ICT based system	Tamil Nadu	http://www.detiace.com/	S Gobinath, Technical Manager DETI@ACE tmdetiace@gmail.com T: 04344-261011,04344-261020 M: 9092327244
76	Bannari Amman Institute of Technology- Technology Business Incubator (BIT-TBI)	Biotechnology in Agro, Industrial and Rural sectors	Tamil Nadu	http://www.bittbi.com/	Dr K BALAKRISHNAN, Principal Scientist & Head Professor balu444@gmail.com T: 04295-221298(Direct), 04295- 226321
77	Centre for Entrepreneurship Development and Incubation (CEDI)	1) IT 2) Electronics	Tamil Nadu	http://www.nitt.edu/home/other/cedi/	Dr. G. Kannabiran, Project Investigator-CEDI cedi@nitt.edu T: 0431-2503003

S.No.	Name	Thrust Area	State	Website Application Process	Contact Details
78	Healthcare Technology Innovation Centre	Medical Devices	Tamil Nadu	https://htic.iitm.ac.in/newsite/	Muthu Singaram, Incharge director@htic.iitm.ac.in M: 9790932420
79	Jeppiaar Engineering College	1) Mechanical 2) Biotechnology 3) Electronics and instrumentation engineering 4) Aeronautical engineering 5) Computer Science 6) IT	Tamil Nadu	http://www.sathyabamauniversity.ac.in/phd/supervisorDetail.php?ref=31055	Dr.M.Sasikumar, Professor In-Charge pmsasi77@gmail.com M: 9094277053
80	Kongu Engineering College - TBI	1) Electronics 2) ICT	Tamil Nadu	http://www.tbi-kec.org/	Mr. P.S.Kannan , Manager 1) tbi-kec@kongu.ac.in 2) tbi_kec@yahoo.com T: 04294 226650 T: 4294 226633 M: 99439 27275
81	Periyar Technology Business Incubator	1) Agro, bio and food technologies 3) Mmanufacturing 2) Herbalproducts	Tamil Nadu	http://www.periyartbi.org/ptbi/index.php	Ms. A.P.Aruna , Chief Executive Officer info@periyartbi.org T: 91+4362-264520, 9944495659 Fax: 91+4362-264660
82	RMK Engineering College - Incubation Center	ITeS	Tamil Nadu	http://www.rmkec.ac.in	Dr. Elwin Chandra Monie, Principal principal@rmkec.ac.in T: 044-33303331/33303330 M: 9443627226
83	Sathyabama UŶiǀersity – Technology Business Incubator	Functional food Bioinformatics Exploitive Micro-biology) Agri-tech & Information Technology 5) Electronics and engineering products	Tamil Nadu	http://sathyabamatbi.in/	Mr Kartik, Manager sathyabamauniversitytbi@gmail.co m M: 9443928573
84	Tiruchirappalli Regional Engineering College - Science and Technology	1) Mechanical 2) IT	Tamil Nadu	http://www.trecstep.com/	R.M.P. Jawahar, Executive Director jawa_ts@yahoo.com T: +91-431-2500085 & 2500697 Fax: +91-431-

S.No.	Name	Thrust Area	State	Website Application Process	Contact Details
85	University of Madras - TBI	1) Manufacturing 2) Healthcare 3) Engineering 4) IT 5) IoT	Tamil Nadu	http://www.unom.ac.in/index.php?route=department/department/about&deptid=69	Dr. G. Gangi Reddy, Managing Director 1)tbi@unom.ac.in 2)tbi_unom@yahoo.com M: 9840597373
86	Vel TeĐh – Technology Incubator	1) Waste Management specifically for E-waste 2) Bio waste 3) Plastics waste 4) 3D Printing 5) Embedded Systems/VLSI 6) Micro Air Vehicles (MAV)/ Robotics 7) Automobile 8) Renewable Energy 9) IT/ITES.	Tamil Nadu	http://www.veltechtbi.com/ http://www.veltechtbi.com/pr	Mr Chandrakumar, Manager vtu@veltechuniv.edu.in T: 044 26840605 Fax: 044 26840605 M: 9600040254
87	Er.Perumal Manimekalai College of Engineering	Brick Manufacturing	Tamil Nadu	http://www.pmctech.org	Dr.S.Chitra, schitra@gmail.com M: 9787009790
88	A.C.College of Engineering & Technology	Civil Engineering	Tamil Nadu	http://www.accet.edu.in	Dr A Sivanantharaja, Associate Professor accet.principal@gmail.com M: 9443919844
89	Agribusiness Development - Tamil Nadu Agricultural University	Agribusiness	Tamil Nadu	https://sites.goole.com/a/tnau.ac.in/dabd/home	Dr R Murugesan, Director business@tnau.ac.in M: 9489056714
90	Coimbatore Innovation and Business Incubator - Kumaraguru College of Technology	1) Electronics (IoT) 2) Technology Services (Industrial/ Engineering Design, Automation)	Tamil Nadu	http://www.forgeforward.in/	Mr. Vishwanathan Sahasranamam , Vice President info@forgeforward.in T: +91 7598446822
91	Dr A.P.J. Abdul Kalam Business Incubator - Vinayaka Mission's Kirupananda Variyar engineering college	1) Biotech 2) Food Processing 3) Electronics equipment design 4) Mechanical equipment design	Tamil Nadu	http://www.vmkvec.ac.in/	Dr. A.Nagappan, Principal principal.vmkvec@vmu.edu.in M: 9842957514, 9362121432

S.No.	Name	Thrust Area	State	Website Application Process	Contact Details
92	K.L.N.College of Engineering	IT	Tamil Nadu	http://www.klnce.edu	1) Dr. A.V. Ramprasad,Head of Institution 2) Dr Pradeep Kanan, Manager Incubation centre principal@klnce.edu 1) M: 9940125238 2) M: 9894770741
93	Nanotech Research Innovation and Incubation Centre (NRIIC) - PSG Institute of Advanced Studies	Nanotechnology	Tamil Nadu	http://www.psgias.ac.in/?p age_id=5150	1) Prof. KK Venkataraman, Prof In charge 2) Dr. K.Suresh Kumar, Manager 1) kkv@psgias.ac.in 2) step@psgtech.edu 1) T: +91-422-4344000 extn 4323 M: 9500950311 Fax: +91-422-2573833 2) T: +91-422-363300, 4363301 Fax: +91-422-2573833
94	PSG-STEP	1) Mechanical 2) IT 3) Nano Technology 4) Electronics	Tamil Nadu	http://www.psgstep.org/	1) Prof. KK Venkataraman, Prof In charge 2) Dr. K.Suresh Kumar, Manager step@psgtech.edu kv@psgias.ac.in T: +91-422-4363300, 4363301 Fax: +91-2573833
95	Sri Ramakrishna Engineering College	ICT	Tamil Nadu	http://www.srec.ac.in	Dr Metilda, Cordinator metilda@srec.ac.in M: 9442952055
96	St. Peter`s Engineering College - TBI	1) Refrigeration 2) Cold chain 3) IT 4) Mechanical	Tamil Nadu	http://www.spectbi.com/	Dr. L. Mahesh Kumar , CEO info@spectbi.com T: [illegible] M: +91-9840024189
97	Technology Business Incubator Centre for Biotechnology - Anna University	Biotechnology	Tamil Nadu	https://www.annauniv.edu/ct dt/contents/technology/biotech.html	Dr. S. Meenakshisundara m, Business Manager meenakshi@annauniv.edu T: 044-22358363; 22350772,22358365 M: 98403 48173 Fax: 044 22350299

S.No.	Name	Thrust Area	State	Website Application Process	Contact Details
98	VIT Technology Business Incubator	1) Bio- Technology based industries 2) Environmentally friendly solutions and products for Leather industries 3) Automotive / Mechanical Engineering sector related products and services 4) Information Technology Products and Solutions	Tamil Nadu	http://www.vittbi.com/#/	Mr A. Balachandran, General Manager vittbi@vit.ac.in T: +91 416 224 3097
99	National Engineering College- Business Incubator	1) Manufacturing 2) Energy 3) Electronics and Trading 4) Agriculture	Tamil Nadu	http://www.nec.edu.in	Dr.S.Shanmugavel, Principal principal@nec.edu.in T: 04632-222502,230227
100	IITM's Rural Technology Business Incubator	1) Rural/Underserved societal segments 2) ICT (Information and Communication Technologies)	Tamil Nadu	http://www.rtbi.in/	Dr. Tamaswati Ghosh , CEO tamaswati@incubation.iitm.ac.in T: +91 44 66469872
102	IIT Madras- Incubation Cell	Agnostic	Tamil Nadu	http://www.incubation.iitm.a c.in/home	Dr. Tamaswati Ghosh , CEO 1)office@incubation.iitm.ac.in 2)tamaswati@incubation.iitm.ac.in T: +91 44 66469869
103	Villgro Innovations Foundation	1) Agriculture 2) Education 3) Energy Health	Tamil Nadu	http://www.villgro.org/	Ms Chaarvi Badani, EA to President info@villgro.org T: 044 66630400
104	IIT Madras Research Park	Agnostic	Tamil Nadu	http://respark.iitm.ac.in/	Dr. Tamaswati Ghosh , CEO tamaswati@incubation.iitm.ac.in T: +91 44 66469872
105	Association for Innovation Development of Entrepreneursh in Agriculture (A- IDEA) - NAARM	Agriculture	Telangana	http://www.aidea.naarm.org.i n	Mr Vijay Nadimti, COO naarmtbi.aidea@gmail.com T: +91 40 24581427 M: +91 99899 28773 Fax: +91-40 24015912

S.No.	Name	Thrust Area	State	Website Application Process	Contact Details
106	Centre for Technology Business Incubator (BITS Pilani)	1) Embedded systems 2) VLSI designs	Telangana	http://www.bits-pilani.ac.in/pilani/technologybusiness/TechnologyBusinessIncubator http://www.bits- pilani.ac.in/Uploads/MicroMo dule/2011-12-17--0-17-59- 528_Application_Form_bitsweb.pdf	Dr. Anu Gupta, Program Manager anug@pilani.bits-pilani.ac.in T: +91-1596-51-5694, 9828125263 Fax: 91 01596 244183
107	International Institute of Information Hyderabad (IIIT)- Foundation	1) Computer vision 2) Natural languages 3) Speech recognition 4) ARNR & gaming domain	Telangana	http://iiith.org/index.html	Mr. Tom Thomas, COO 1)http://iiith.org/contact.html 2)tom@iiith.org T: 040 6653 1211
108	T-HUB	Agnostic	Telangana	http://www.t-hub.co/	Mr. Jay Krishnan, jay.krishnan@t-hub.co M: 9945163858
109	NASSCOM 10K Warehouse Telengana	T and enabled Technologies	Telangana	http://10000startups.com/10k-program/	Vijay Kumar Bawra, Manager vijay@nasscom.in M: 9831524485
110	Technology Business Incubator (University of Hyderabad)	1) Pharma 2) Biotechnology 3) Renewable energy	Telangana	http://www.uohyd.ac.in/index.php/about-uoh?layout=edit&id=819	Dr. B.S. Rama Krishna, Administrative Co- ordinator 1)admintbi@uohyd.ernet.in 2)dr.ramakrishna@gmail.com T: 040-23137551 M: 09885040914 Fax. 040-23011091
111	ALEAP- India	Eco Friendly products	Telangana	http://www.aleap.org	Mr A.Ajay Babu- Manager -projects /Mrs. Lakshmi- Manager ,ALEAP Incubators aleap93@gmail.com 8978988005
112	Dayalbagh Educational Institute	1) Textile 2) Apparel 3) Solar 4) Dairy 5) ElectricalVehicles	Uttar Pradesh	http://www.dei.ac.in/dei/	Prof.Prem Kumar Kalra , Director 1)drpremkalra@gmail.com 2) director@dei.ac.in M:9458553555 T: 0562-6548399/2801545 Fax: 0562-2801226

S.No.	Name	Thrust Area	State	Website Application Process	Contact Details
113	Doon College of Engineering & Technology	1) Technology 2) Pharma 3) Agriculture	Uttar Pradesh	http://www. dooncollege.in fo	Praveen Chaudhary, Additional Manager info@dooncollege.in M: 9412007104
114	IIITA Info Communication Incubation Centre (IIIC)	1) IT 2) Biomedical engineering	Uttar Pradesh		Prof. Sudip Sanyal, Professor ssanyal@ iiita.ac.in M: +91-9415235180
115	IIMT College of Pharmacy	1) Pharma Industries 2) Healthcare	Uttar Pradesh	http://www.iimtindia. net	Dr. Mallikarjun BP, Principal iimtgnoida@iimtindia.net Fax: 0120-2323860 M: 9990888627 T: 0120-2475025
116	Noida Institute of Engineering Technology Technology Business Incubator	1) ICT 2) Biotechnology	Uttar Pradesh	http://niet.co.in/	Mr. S.R.Mustafa, General Manager srmustafa@niet.co.in M: 7838204509
117	Amity Innovation Incubator		Uttar Pradesh	http://www.amity. edu/aii	Tel No. 0120-4659000, 4392243
118	Process Cum Product Development center	1) Sporting goods 2) Handicraft 3) Leather	Uttar Pradesh	http://www. ppdcmeerut.com	1) Mr.Sunil Gupta, Principal Director 2) VK Singh, Person in Charge info@ ppdcmeerut.com M: 9410269600 (VK Singh) T: 0121-2511779
119	SIDBI Innovation and Incubation Centre	1) Technology 2) Engineering and other interdisciplinary areas	Uttar Pradesh	http://www.iitk.ac.in/ siic/d/	1) Dr. Sameer Khandekar, Co ordinator 2) S. Sudha, Manager 3) Ravi Pandey, Senior Associate (Patents &Technology Transfer) 1)samkhan@iitk.ac.in 2) siic@iitk.ac.in 3) rpandey@iitk.ac.in 1) T: +91 - 512 - 2596646; +91 - 512 - 2596606 2) T: +91 - 512 - 2597057 3) T: +91 - 512 - 6596178
120	TBI- KIET (Krishna Institute of Engineering & Technology)	1) ICT 2) Mechanical 3) Electronics	Uttar Pradesh	http://www.tbi-kiet.in/	Prof. Satyendra Kumar , General Manager tbikiet@gmail.com M: 9999017712

S.No.	Name	Thrust Area	State	Website Application Process	Contact Details
121	IAN Mentoring & Incubation Services	Agnostic	Uttar Pradesh	http://delhi.tie.org/ian-incubator-powered-by-tie-delhi-ncr/	Ms. Archana Jahagirdar archana@tienewdelhi.org
122	TIDES Incubation Centre, IIT Roorkee (Technology Innovation and Development of Entrepreneurship Support Centre, IIT Roorkee)	1) Information & Communication Technology (ICT) 2) Bio Technology 3) Manufacturing and Engineering 4) Micro and Nano Electronics 5) Water, Sanitation and Solid Waste Management 6) Housing-Urban and Rural 7) New Materials including nano Materials 8) Building Materails / Construction Technology 9) Renewable Energy	Uttarakhand		Dr. Sanjeev Manhas, Associate Professor, Faculty In-Charge 1) smanhas333@gmail.com 2) tides@iitr.ac.in T: 01332-285147, 9760511744
123	IIM Calcutta Innovation Park	1) Social Enterprises 2) MSME	West Bengal	http://www.iimcip.org/	Mr. Subhrangshu Sanyal , CEO subhrangshu.sanyal @iimcip.org T: +91 33 2467 8313 (Extn 576/577/541)
124	Technology Incubation and Entrepreneurship Society (TIETS)	Agnostic	West Bengal	http://www.step-iit.org/about_tiets.html	1) Dr. Satyahari Dey, Professor, Managing Director 2) Prof. Siddhartha Das, Executive Adviser 3) Mr. Subhash Chandra Santra , General Manager mdstep@hijli.iitkgp.ernet.in 1) T: 91-3222-283760 or- 282247 (Office), - 277097 (Res.), -281366 (laboratory) Fax: 91-3222-255303 or-278707 2) T: +91-3222-283256 ,+91-3222- 281091 3) T: 03222-281091
125	Nasscom 10000 Startups Warehouse Kolkata	Technology	West Bengal	http://10000startups.com	Ravi Ranjan, Manager ravi@nasscom.in M: 9874443395
126	Tagore Centre for Green Technology Business Incubation	Agnostic	West Bengal	http://tcgtbi.iiests.ac.in/index	Pratho Chakraborty, General Manager 1) mksanyal@tcgtbi.org 2) hodhrm@becs.ac.in T: 033-26681073 Fax: 033-26689925 M: 09831352950

S.No.	Name	Thrust Area	State	Website Application Process	Contact Details
127	Ekta Incubation Centre	1) Biotechnology 2) IT	West Bengal	http://www.technologyembryo.com/	Mr. Alok Ray, Director, EKTA- TBI info@technologyembryo.com ektaincubationcentre@gmail.com T: (033) 2367 3978
128	Society for Innovation & Entrepreneurship (SINE), IIT Bombay	Technology	Maharashtra	www.sineiitb.org	

Annexure-VII

INVESTMENT FORMAT SAMPLE FOR INNOVATORS

APPLICATION OF ENTREPRENEUR / INCUBATEE

A. GENERAL INFORMATION

a. Applicant's Name:

b. Age (Years): DOB (dd/mm/yyyy):

c. Highest Qualification:

d. Address of the Applicant:

e. Phone/Mobile (Mandatory):

f. Email (Mandatory):

B. STARTUP DETAILS

a. Name of Startup:

b. Address of Startup:

c. Have you registered your startup? (Please tick (ü) the applicable option)

YES ____ NO___

d. What is the legal status of the entity you have formed? (Please tick (ü) the applicable option)

Sole Proprietorship		Partnership		Limited Liability Partnership		Private Limited Company	

e. Founding Year:

f. Website:

g. Founding Team details with shareholding pattern:

Name	*Position*	*Responsibilities*	*Phone/ Mobile*	*% Share of company*	*Email*

C. TECHNOLOGY

a. Which Focus Area does your product/service belongs?

b. Where are you in your product development / path to market process?

- Idea
- Prototype
- MVP
- Beta customers
- Commercially operational

c. Please explain your technology in brief: Technology Description: Intended product or service (Define scope carefully). For a process, one needs to think of the product that will result.

(i) Explain your Technology/ Innovation.

(ii) Explain USP (Unique Selling Proposition) of your product/service? Please focus on "benefits" rather than "features" of the technology.

(iii) What does it do and what problem it solves? (Use simple and non-technical terms).

(iv) Does your technology have any intellectual property protection? If not, what is your plan to protect your technology from competitors?

d. Commercial and existing technology landscape

(i) What is the competitive advantage of your technology with respect to existing products/services/technologies in the market today?

e. How long will it take for this innovation/technology is completely developed and ready for the market? (Please tick (✓) the applicable option)

- Already in active use now
- 1-6 months
- 7-12 months
- 12+ months

f. What are the key risks/ barriers that you face in bringing your technology to the market? Please explain (Technical/ manufacturability, Freedom to operate, Regulatory, compliance, finance, scale of investment barriers, team etc.)

D. BUSINESS AND INVESTMENT PLAN:

(a) Describe your current traction: mention relevant metrics such as: Monthly growth, Nos. of customers, MAU (Monthly active users), DAU (Daily active User), etc.

(b) Describe your current revenue model?

(c) Please explain your business plan and investment plan for the next one year.

(i) Market assessment: Size and growth. Please give an estimate of your (current) target market segments(B2B: B2C: B2B2C)

(ii) Competitive advantage of your product/ technology over your competitors.

(iii) Explain your current marketing/ sales/ distribution strategy.

(d) Please explain your current investment plan for the next one year. Please high-light immediate milestone for the next 6 months;

(i) Develop a cash flow statement considering your current and future needs.

(ii) What are your revenue projections for the next one year?

(iii) What is your funding requirement? Provide a brief milestone and timelines for the next one year in the below format.

(iii) Fund Utilization from grant. *

E. OTHER RELATED INFORMATION

Please include any further information that you wish to communicate to us to help us in judging your application

Declaration:

I ________ hereby certify that the information furnished in the application form from Item A-E is true, complete and to the best of my knowledge.

Date and Place:

Date –

Signature of the Applicant

References

Allen, K. R. (2006) Launching New Ventures: An Entrepreneurial Approach, 4th edition, Houghton Mifflin.

Amabile, T. M. (1996) 'Creativity and innovation in organisations', HBS note 9-396-239

Baron, R. A., & Shane, S. A. (2006) Entrepreneurship: A Process Perspective, Mason, Ohio: Thomson / South-Western.

Barquet, Ana Paula B., et al. "Business model elements for product-service system". Functional Thinking for Value Creation. Springer Berlin Heidelberg, 2011. 332–337: They stated that "The Canvas business model was applied and tested in many organizations (eg IBM and Ericsson), being successfully used to easily describe and manipulate business models to create new strategic alternatives."

Bhave, M. P. (1994)' A process model of entrepreneurial venture creation', Journal of Business Venturing 9(3): 223-242.

Bower, J. L. & C. M. Christensen (1995) 'Disruptive technologies: Catching the wave', Harvard Business Review 73: 43-53.

Brush, C. G., P. G. Greene, & Hart, M.M. (2001) 'From initial idea to unique advantage: the entrepreneurial challenge of constructing a resource base', Academy of Management Executive, 15(1): 64-80

Bygrave, W., Zacharakis, A. (2008) Entrepreneurship, Hoboken, NJ: Wiley.

Carland, J. W., F. Hoy, W. Bolton, & J.A.C. Carland (1984) 'Differentiating entrepreneurs from small business owners: A conceptualisation, The Academy of Management Review 9(2): 354-359.

Christensen, C. M. & Raynor, M. E. (2003) The Innovator's Solution, Boston, MA: Harvard Business School Press.

David E. Rumelhart, 'Schemata: the building blocks of cognition'. In: R.J. Spiro et al. (eds) Theoretical Issues in Reading Comprehension, Hillsdale, NJ: Lawrence Erlbaum, 1980

De Reuver, Mark, Harry Bouwman, and Timber Haaker. "Business model roadmapping: A practical approach to come from an existing to a desired business model". International Journal of Innovation Management 17.01 (2013): They described the business model canvas as the "Most prominent... popular tool that makes it simple for practitioners to design business models in a creative session."

Drucker, P. F. (1998) 'The discipline of innovation, Harvard Business Review 76(6): 149-157.

Frederick, H.H., Kuratko, D. F. & R. M. Hodgetts (2007) Entrepreneurship: Theory, Process, Practice, Thomson South-Western.

Freeman, J. & Engel, J. S. (2007) 'Models of innovation: Startups and mature corporations, California Management Review, 50: 94-119.

Gartner, W. B. (1985) 'A conceptual framework for describing the phenomenon of new venture creation, The Academy of Management Review 10(4): 696-706.

Gartner. W.B. (2007). "Entrepreneurial narrative and a science of the imagination". Journal of Business Venturing. 22 (5): 613–627. doi:10.1016/j.jbusvent.2006.10.003

Ghemawat, P. (2007) Redefining Global Strategy. Boston, MA: Harvard Business School Publishing.

Global Entrepreneurship Monitor (multiple years), London Business School / Babson. Information on GEM is available from http://www.gemconsortium.org

Gompers, P. A. (1998) 'A Note on Valuation in Entrepreneurial Ventures', HBS note 9-298-082.

Hamel, G. (1999) 'Bringing Silicon Valley inside', Harvard Business Review, September: 70-84.

Hamermesh, R. G., & P. W. Marshall (2002) 'Note on Business Model Analysis for the Entrepreneur,' Harvard Business School Research Notes: 048.

Hammermesh, R.G. (2007) 'The Entrepreneurial Manager, 2008 Winter Term: Course Outline and Syllabus', HBS note 9-801-325.

https://krishijagran.com/featured/who-is-going-to-fund-your-agristartup/

Hisrich, R. D., Peters, M. P., & Shepherd, D. A. (2008) Entrepreneurship (7th ed.). Boston: McGraw Hill-Irwin.

https://startuptalky.com/agriculture-startups-in-india/

Kaplan, J. M. & Warren, A. C. (2007) Patterns of Entrepreneurship (2nd ed.). Hoboken, NJ: John Wiley & Sons.

Kawasaki, G. (2004) The Art of the Start. New York: Portfolio.

Knight, G. A. & Cavusgil, S. T. (2004) 'Innovation, organisational capabilities, and the born-global firm', Journal of International Business Studies, 35: 124-141.

Ko, S. & J. E. Butler (2007) 'Creativity: A key link to entrepreneurial behavior', Business Horizons 50(5): 365-372.

Kogut, B. & Kulatilaka, N. (1994) 'Options thinking and platform investments: Investing in opportunity', California Management Review, 36(2): 52-71.

Kogut, B. (1989) 'Research Notes and Communications. A Note on Global Strategies', Strategic Management Journal, 10: 383-389.

Kuemmerle, W. (2005) 'The entrepreneur's path to global expansion', MIT Sloan Management Review, 46(2): 42-49.

Leslie, K. J. & M. P. Michaels (1997) 'The real power of real options', The McKinsey Quarterly (3): 5-22.

Luecke, R. (2004) Entrepreneur's Toolkit: Tools and Techniques to Launch and Grow Your New Business, Boston: Harvard Business School Press, Chapter 5.

Matofska, B., 2015. What is the Sharing Economy? [online] Thepeoplewhoshare.com. Available at: <http://www.thepeoplewhoshare.com/blog/what-is-the-sharingeconomy/> [Accessed 14 March 2022].

Mazzarol, T. (2006) Entrepreneurship & Innovation: A Manager's Perspective, Prahran: Tilde University Press.

McMullen, J. S. & D. A. Shepherd (2006) 'Entrepreneurial action and the role of uncertainty in the theory of the entrepreneur', The Academy of Management Review 31(1): 132-152.

Osterwalder, A., Pigneur, Y. and Clark, T., 2010. Business model generation. [Amsterdam]: Alexander Osterwalder & Yves Pigneur.

Osterwalder, Alexander (2005-11-05). "What is a business model?". business-model-design.blogspot.com. Archived from the original on 2006-12-13. Retrieved 2019-06-19. ... we could define a business model as a simplified description of how a company does business without having to go into the complex details of all its strategy, processes, units, rules, hierarchies, workflows, and systems. However, now that we know that the business model is a simplified representation of how we do business, we still have to decide which elements to describe. A synthesis of literature shows that there are mainly 9 building blocks to help us describe a business model.

Osterwalder, Alexander (2004). The Business Model Ontology: A Proposition In A Design Science Approach (PDF) (Ph.D. thesis). Lausanne: University of Lausanne. OCLC 717647749. See also: Osterwalder, Alexander; Pigneur, Yves; Tucci, Christopher L. (2005). "Clarifying business models: origins, present, and future of the concept". Communications of the Association for Information Systems. 16 (1): 1. doi:10.17705/1CAIS.01601

Oviatt, B. M. & McDougall, P. P. (1995) 'Global start-ups: Entrepreneurs on a worldwide stage', Academy of Management Executive, 9(2): 30-43.

Patriotta, G and Siegel, D (2019). "The Context of Entrepreneurship". Journal of Management Studies. 56 (6): 1194–1196.

PR Newswire, 2015. Business Model Canvas Replaces Traditional Business Plans. [online] Available at:<http://ezproxy.lib.uts.edu.au/login?url=http://search.proquest.com/docview/1653118734?accountid=17095> [Accessed 30 March 2015].

Rennie, M. W. (1993) 'Born global', The McKinsey Quarterly, 4: 45-53.

Rhinesmith, S. H. (1992) 'Global Mindsets for global managers', Training & Development, October: 63-68.

Rich, S. R. & Gumpert, D. E. (1985), 'How to write a winning business plan', Harvard Business Review, 3: 3-8.

Sahlman, W. A. (1996) 'Some Thoughts on Business Plans', HBS note 9-897-l 01.

Sahlman, W. A. (1997) 'How to Write a Great Business Plan', Harvard Business Review, July: 98-108.

Shane, S. & S. Venkataraman (2000) 'The promise of entrepreneurship as a field of research', The Academy of Management Review 25(1): 217-226.

Sandra L. Bem, 'Gender schema theory: A cognitive account of sex typing', Psychological Review, Vol 88(4), p 354-364, Jul 1981

Stevenson, H., H., & Roberts, M. J. (2002), 'New Venture Financing', HBS note 9-802-131.

Thornberry, N. (2001) 'Corporate entrepreneurship: antidote or oxymoron?' European Management Journal, 19: 526-533.

Timmons, J. A. & Spinelli, S. 2007. New Venture Creation (7th edition ed.). Boston: McGraw Hill.

Wickham, P. A. (2006) Strategic Entrepreneurship (4th edition ed.). Harlow, England: FT Prentice Hall.

Wolcott, R. C. & Lippitz, M. J. (2007) 'The four models of corporate entrepreneurship', MIT Sloan Management Review, 49:75-82.

Index